New Directions in the Philosophy of Science

Series Editor

Steven French
Dept Philosophy
Univ Leeds
Leeds, United Kingdom

The philosophy of science is going through exciting times. New and productive relationships are being sought with the history of science. Illuminating and innovative comparisons are being developed between the philosophy of science and the philosophy of art. The role of mathematics in science is being opened up to renewed scrutiny in the light of original case studies. The philosophies of particular sciences are both drawing on and feeding into new work in metaphysics and the relationships between science, metaphysics and the philosophy of science in general are being re-examined and reconfigured. The intention behind this new series from Palgrave-Macmillan is to offer a new, dedicated, publishing forum for the kind of exciting new work in the philosophy of science that embraces novel directions and fresh perspectives. To this end, our aim is to publish books that address issues in the philosophy of science in the light of these new developments, including those that attempt to initiate a dialogue between various perspectives, offer constructive and insightful critiques, or bring new areas of science under philosophical scrutiny. The members of the editorial board of this series are: Otavio Bueno, Philosophy, University of Miami (USA) Anjan Chakravartty, University of Notre Dame (USA) Hasok Chang, History and Philosophy of Science, Cambridge (UK) Steven French, Philosophy, University of Leeds (UK) series editor Roman Frigg, Philosophy, LSE (UK) James Ladyman, Philosophy, University of Bristol (UK) Michela Massimi, Science and Technology Studies, UCL (UK) Sandra Mitchell, History and Philosophy of Science, University of Pittsburgh (USA) Stathis Psillos, Philosophy and History of Science, University of Athens (Greece) Forthcoming titles include: Sorin Bangu, Mathematics in Science: A Philosophical Perspective Gabriele Contessa, Scientific Models and Representation Michael Shaffer, Counterfactuals and Scientific Realism Initial proposals (of no more than 1000 words) can be sent to Steven French at s.r.d.french@leeds.ac.uk.

More information about this series at
http://www.springer.com/series/14743

Vassilis Livanios

Science in Metaphysics

Exploring the Metaphysics of Properties and Laws

Vassilis Livanios
University of Cyprus
Nicosia, Cyprus

New Directions in the Philosophy of Science
ISBN 978-3-319-41290-0 ISBN 978-3-319-41291-7 (eBook)
DOI 10.1007/978-3-319-41291-7

Library of Congress Control Number: 2016958225

Printed on acid-free paper

This Palgrave Macmillan imprint is published by Springer Nature
The registered company is Springer International Publishing AG
The registered company address is:Gewerbestrasse 11, 6330 Cham, Switzerland

Series Editor's Preface

Recent years have seen a flourishing of what has been called 'the metaphysics of science'. In part this is the result of certain challenges to metaphysics as currently practised, which have focused on its apparent lack of engagement with modern science. A quick scan of the relevant literature will reveal numerous works on monism, dispositions, metaphysical 'simples', gunk and other such exotica but which either deploy caricature-like examples from physics, say, or make no contact with current scientific developments whatsoever. When so challenged, metaphysicians will typically respond that they are more concerned with charting the vast seas of the possible than with accommodating the mundane features of the actual world, but that raises the obvious question: if their metaphysics of the actual fails to make contact with modern science, why should we pay attention to their claims about the possible?!

Fortunately, Livanios' aim is to meet this challenge by presenting a 'science-informed metaphysics' that takes seriously not just the results of current physical science but also what we know of how that science works. His carefully constructed and detailed analysis begins with one of the most prominent metaphysical claims—namely, that all physical properties can be understood in terms of dispositions that manifest in certain ways when subjected to certain stimuli—and shows how it is undermined by General Relativity, for example, which suggests that spatio-temporal relations are in fact categorical. This then feeds into his defence of 'Categorical Monism' which insists that all fundamental natural properties are categorical and non-dispositional, a defence that also draws on certain much discussed features of Quantum Field Theory.

In subsequent chapters, Livanios tackles some well-known objections to this view and, further, argues that as a metaphysical account of fundamental properties, it is only contingently true. In other words, although in this world the properties of physics are categorical, there are possible worlds in which they are dispositional. This is a provocative claim which then shapes his consideration of the metaphysics of scientific laws. Here he rules out currently popular dispositional accounts, on the grounds that they cannot accommodate conservation laws and associated symmetries and defends a form of 'nomic realism', which maintains that certain law-like relations, at least, actually exist, where, again, this should be understood as a contingent matter.

Thus, Livanios' book not only deals with a number of much discussed issues that sit at the heart of the metaphysics of science, issues to do with the nature of properties and of laws, but also exemplifies beautifully and engagingly how a 'science-informed metaphysics' should proceed. It subjects a range of alternative perspectives to the detailed and nuanced consideration that has been lacking in many such accounts and also advances a radical new approach to the metaphysics of scientific laws, namely 'selective nomic realism'. For both these sets of reasons, then, *Science in Metaphysics* is a very worthy addition to the *New Directions in Philosophy of Science* series, and I am sure it will not only provoke considerable discussion but also open up new avenues of research at the intersection of metaphysics and the philosophy of science.

Series Editor Steven French,
Professor of Philosophy of Science,
University of Leeds.

Contents

Introduction

Surveying the introductions of a number of recent essays on metaphysics (see, for instance, Loux (1998), Sider (2001), Lowe (2002), Heil (2003)) one may find that the authors embark on their work with attempts to defend metaphysics as a discipline worth pursuing. This is an understandable attitude given the recent resurgence of an anti-metaphysical stance expressed in various views concerning whether metaphysics has anything to say about the structure of reality at all (for a recent collection on that meta-metaphysical issue, see Chalmers et al. (2009)). Unlike those metaphysicians, I do not feel obliged to do that; in my view, metaphysics needs no more defence than philosophy itself in general. So, I begin unhesitatingly (and unblushingly!) by claiming that the present work is intended to be an essay falling under the broad category of *science-informed* metaphysics. In particular, it aims to be an essay in metaphysics of *physical* science. This means (inter alia) that throughout this work I have taken seriously both the findings of mature, fundamental physical theories *and* the scientific practice. The relationship between science and metaphysics is a subtle one; therefore, it is somewhat controversial how this 'taking seriously' clause should be cashed out. To start off, however, it suffices to point out that my commitment to a science-informed manner of doing metaphysics compels me to give extra weight to arguments stemming from (or simply inspired by) the findings and practice of contemporary (physical) science, and to ensure that my metaphysical suggestions are (at least) compatible with the latter. In opposition to Ladyman and Ross (2007), I do not think that the above commitments put *severe* constraints either on what the proper manner of doing metaphysics should be or on the set of metaphysical questions worth examining. I certainly agree that metaphysics *may* have an ancillary role *vis-à-vis* philosophy of science. As French and McKenzie (2012) argue, metaphysics can provide a 'toolbox' of concepts which might help the philosopher of science to articulate a metaphysical background for contemporary theories and to understand through it, at least, what the world would be like if our theories were true. Nevertheless, I think that metaphysics has its own distinctive and worthy-of-examination problems. So here I shall talk about *metaphysics*. The issues under consideration in this book concern *purely* metaphysical problems related to the nature of fundamental properties and laws. In a certain sense, then, what I am doing in the work at hand is *science in metaphysics*. In particular, I propose to explore the various ways in which modern physical theory can throw light on the metaphysical problems in question.

As I see it, the science-informed route of metaphysical investigation is motivated by a broadly *realistic* view about science as the most reliable tool for investigating and providing approximately true descriptions of the world. Nevertheless, the various interrelationships between science and metaphysics do not *compel* us to think of the latter as an investigation of the most general categories of being, their features and the relations among them (horizontal relations and vertical ones, concerning grounding and metaphysical dependence). We cannot rule out the option that metaphysics concerns the *concepts* we use. But, in that case too, being science-informed provides us with the best reason to think that those concepts are intended to capture the mind-independent structure of the world.

Even if metaphysics concerns the structure of the world, the possibility that some sort of a priori reasoning can reveal aspects of the latter cannot be ruled out. Perhaps, there exist certain purely metaphysical problems which do concern mind-independent reality but have failed up until now to be science-informed. In that case it is only our intuitions that may ground our judgement about those problems. Besides, our intuitions (which are mainly related to those sorts of a priori reasoning), though relative and often unreliable (science has shown that!), are reality-informed in a broad sense; so there exists a possibility of a kind of scale-invariance (in some respects) providing intuitions with a methodological reliability as far as the investigation of *some* aspects of the fundamental structure of the world. (Perhaps, for instance, some of the formal features of general metaphysical relations that intuitions reveal hold not only at the non-fundamental but also at the fundamental level; consider, for example, the justification for the exclusion of certain 'pathological' models of General Relativity which violate some basic intuitions about causation.) Hence, as far as I can see, the science-informed manner of metaphysical investigation does not exclude a priori reasoning about metaphysical problems.

Setting aside issues methodological, this book aims to throw some light on a purely metaphysical issue concerning the nature[1] of properties and relations. Metaphysics of properties is a highly articulated branch of philosophy investigating a multitude of interesting issues. A survey would reveal at least three main categories of issues under scrutiny. The first focuses on the very *genuine* existence of properties. Are there only concrete particulars (and perhaps their resemblances and classes thereof, as extreme nominalists claim) or does the ontological inventory of the world also (or, according to some, exclusively) include properties and/or relations (Property-Realism[2])? The second category concerns the proper ontological account of genuinely existing properties. Are the latter universals (Universalism) or particulars (Tropism, aka Moderate Nominalism)?

Though both of the above topics are extremely significant (from the metaphysical point of view), they fall beyond the scope of this book which is focused on the last (and very popular nowadays) category of issues concerning the dispositional or categorical nature of properties. Are all

[1] Throughout this book, my use of the term 'nature' is (unless explicitly mentioned) a bit idiosyncratic: First, because it has a restrictive scope (referring, as we shall see, only to the dispositional and/or categorical character of properties); and, second, because the nature of properties (as here intended) is not necessarily related to the possession of essential (or even metaphysically necessary) ontological features.

[2] The term Property-Realism often refers to Universalism, that is, the thesis according to which at least all the sparse, natural properties are universals. I, instead, use the term to refer more broadly to the view which allows the existence of a non-empty ontological category of genuinely existing properties and relations.

properties of the same nature or not? Proponents of the latter option (dualists) argue that some properties are dispositional, while others are categorical. Defenders of the former option are divided; dispositional and categorical monists claim that all properties are of the same nature, either dispositional or categorical, respectively. Identity theorists (and double-aspect/dual-sided theorists) hold that all properties are *both* dispositional and categorical, while neutral monists remain agnostic on that issue. In this book, I join the monistic camp and defend a sui generis Categorical Monism about the fundamental properties and relations of our world.

Every work has its own presuppositions (books on metaphysics are of course no exceptions to the rule) and the sooner they are laid out the better. So, to begin with, the metaphysical investigations in this essay presuppose a *realistic* approach to fundamental, first-order properties (at least to the extent that they are absolute determinates) without being explicitly committed to either Universalism or Tropism. I also presuppose that, in the actual world, there are ontologically genuine relations (in the sense that they do not supervene on monadic properties). Most of what I have to say in this work applies *mutatis mutandis* to both (fundamental) monadic properties and relations. So, when I refer in the sequel to properties, I mean (except in cases where I explicitly deny it) both monadic and polyadic ones. Furthermore, I should make clear that the whole discussion in what follows does not concern mathematical, logical or mere Cambridge properties. It is about the natural (sparse, in Lewis' terminology) properties that carve nature at its joints. Especially, it is largely about the *fundamental* features of the actual world. There is a relatively broad consensus that paradigmatic examples of currently accepted fundamental properties are electric charge, inertial–gravitational mass, spin,[3] colour (of quarks), and so on, while instances of fundamental relations are most probably the spatiotemporal relations. Hence, though reference will occasionally be made to non-fundamental properties of medium-sized ordinary objects, almost all the arguments supporting the theses in this book refer to the above features, which are what contemporary

[3] These are the so-called *state-independent* properties, because they take constant, determinate values, independently of the state of the particle (or the system). For instance, an electron always has rest mass $m=9 \cdot 10^{-31}$ kg, charge $q=-1{,}6 \cdot 10^{-19}$C and spin $s=1/2$.

physical science acknowledges as fundamental properties and relations of the elementary entities of the world.

The structure of the book is as follows: I embark, in the following two chapters, on the task of examining all current metaphysical accounts of the nature of fundamental features in order to find the one fitting the actual world. The discussion begins in Chap. 1 with Dispositional Monism, according to which all actual fundamental properties and relations are dispositional. My argument against Dispositional Monism is direct; I argue that the spatiotemporal metric relations are genuine *purely categorical* features of the actual world. To be more precise, after presenting my preferred *truthmaking*-criterion for the dispositional/categorical distinction, I proceed to show—contrary to the arguments of Bartels (2013), Esfeld and Sachse (2011) and, especially, Bird (2007)—that the most mature and successful theory of spacetime we currently have (i.e., the General Theory of Relativity) strongly suggests that the spatiotemporal metric relations are categorical.

Chapter 2 is devoted to the examination of the prospects of two other metaphysical accounts of the nature of properties, that is, Identity Theory and Neutral Monism. I argue that the 'big bad bug' (to use Lewis' term) for Identity Theory is how to understand the surprising triple identity between a property and its two apparently contradictory aspects. Neutral Monism (according to which the dispositional/categorical distinction concerns the *predicates* we use to describe the actual fundamental properties and not the properties themselves) is the second view discussed in this chapter. My argumentation inter alia highlights the fact that the most serious objection to Neutral Monism is that it compels property-realists to adopt an *agnostic* stance towards the issue concerning what the proper nature of fundamental properties should be if it were representable by two different epistemic descriptions. For it is difficult to understand how a metaphysician who takes the debate between the two accounts seriously (and I am sure that the overwhelming majority of property theorists does) may nevertheless refrain from investigating its ontological grounds in the nature of the fundamental properties.

The arguments of Chaps. 1 and 2, if sound, leave us with two options: to embrace either Categorical Monism or Property-Dualism. Until recently (see, for instance, Livanios (2012a)), I was convinced that the latter is most probably the correct metaphysical account of the nature of actual fundamental properties. However, paying closer attention to the arguments that dispositional realists advance against Categorical Monism, I gradually lost my (relative) confidence in the actual truth of Property-Dualism. Eventually, I changed my mind when I found a scientifically based reason which convinced me that all fundamental properties of the actual world are categorical. The argument in defence of Categorical Monism is presented in Chap. 3 and is inspired by the renormalisation methods used in Quantum Field Theories.

The next two chapters are devoted to addressing some apparently serious difficulties for Categorical Monism. In Chap. 4, I discuss the principal objection that the nature of categorical properties commits the categorical monist to accept counter-intuitive modal scenarios. First, I present the two prominent rival views about the grounds of the *de re* modal representation of properties—the Dispositionalist View (DV) and the Radical Non-Dispositionalist View (RNDV)—and show that the latter, which is associated with categorical features, is a much richer account than it is commonly assumed. Then, I defend RNDV against what I call the Permutation Argument. According to the latter, the adoption of RNDV leads to the counter-intuitive view that certain possible worlds (generated by permutations of fundamental properties among the causal/nomic roles they occupy) are distinct.

Chapter 5 examines further objections to Categorical Monism. The problems I discuss in this chapter include the putative incapability of Categorical Monism to provide adequate truthmakers for unmanifested dispositions, as well as various arguments for the existence of genuine fundamental dispositional properties 'emerging' both from the scientific way of characterising properties and scientific practice.

In Chap. 6, I suggest that the very methodology of science-sensitive metaphysics indicates that the true metaphysical account of the nature of actual fundamental properties (in my view, Categorical Monism) is only *contingently* true. A corollary of that is what I call the thesis of ConD/C; that is, the view according to which the fundamental features of our world are *contingently* categorical and, consequently, there are possible worlds in which they (or their counterparts) are dispositional. The best part of the chapter is devoted to addressing objections to this unorthodox view, ranging from accusations of unintelligibility and non-viability to explanatory deficiency.

In the next two chapters of the book, I turn my attention to the metaphysics of laws of nature. This is a natural move to make since the metaphysical accounts of properties most often come as a package deal with a corresponding account of laws. Chapter 7 discusses the general existential question about laws. In view of my commitment to the metaphysically contingent character of categoricality and dispositionality, my goal in this chapter is not only to answer the question in the actual world but also in worlds where Dispositional Monism or Property-Dualism is true. Nomic Realism is the view I wish to defend here, and to this end I argue against a specific account of laws (the one I call Dispositional Essentialist Account of Laws—DEAL) certain versions of which threaten Nomic Realism both in dispositional monistic and in property dualistic possible worlds. After a lengthy discussion, I conclude that conservation laws (and the corresponding symmetries), as well as the presence of fundamental constants in laws, strongly suggest that DEAL is an inadequate account.

Chapter 8 examines two important metaphysical issues concerning features of nomic relations and laws. Starting from their modal status, an argument for the robust metaphysical contingency of laws is put forward. Then, I introduce the notion of a *hybrid* nomic relation and suggest that *mixed* nomic relations (i.e., those which have both dispositional and categorical relata) are neither external nor internal but belong instead in that third kind. Finally, Chap. 9 includes some concluding remarks.

Some of the material presented in this book has already been published in the following papers:

'The "Constant" Threat to the Dispositional Essentialist Conception of Laws', *Metaphysica*, 15(1), 2014, 129–155.

'Radical Non-Dispositionalism and the Permutation Problem', *Axiomathes*, 24 (1), 2014, 45–61.

'Exploring the Metaphysics of Nomic Relations', *Acta Analytica*, 27, 2012, 247–264.

'On the Possibility of Contingently Dispositional Properties', *Abstracta*, 6 (1), 2010, 3–17.

'Symmetries, Dispositions and Essences', *Philosophical Studies*, 148 (2), 2010, 295–305.

'Bird and the Dispositional Essentialist Account of Spatiotemporal Relations', *Journal for General Philosophy of Science*, 39(2), 2008, 383–394.

I would like to thank the publishers (Springer Science + Business Media and Walter de Gruyter GmbH) for their permission to use material from the above papers.

Part of the research for this book was carried out while I was benefiting from the Research Funding Program: THALIS–UOA-70/3/11604, MIS 375791 (Operational Program 'Education and Lifelong Learning' of the National Strategic Reference Framework (NSRF)) co-financed by the European Union (European Social Fund—ESF) and Greek national funds.

A number of people who attended the conferences and workshops in which part of the material of this work was previously presented helped me to avoid errors and improve my arguments. I am indebted to all of them. I am grateful to Steven French for his support, and Brendan George and Grace Jackson (from Palgrave Macmillan) for their help. I am especially grateful to Michael Esfeld, Matteo Morganti and two anonymous reviewers for their detailed comments and suggestions. I would also like to thank Demetra Christopoulou, Filippos Georgiadis, Pandora Hatzidaki, Stavros Ioannidis and Vassilis Sakelariou (participants of a reading group I organised in the University of Athens), who read and commented on an earlier draft of the book. Especially thanks to Vassilis Sakelariou for the language editing. I owe a great debt to my teacher and friend Stathis Psillos for his long-standing help and encouragement. Finally, I would like to express my gratitude to my wife Alexandra, without whose love, patience and support this book would never have been finished.

1

Against Dispositional Monism

1.1 Preliminaries: Two Distinct Issues

Contemporary metaphysics of properties and laws is dominated by two worldviews. According to the neo-Humean view, fundamental properties and relations in the actual world are *intrinsically* causally inert, while their transworld identity (or, more generally, their *de re* modal representation) is completely independent of their actual causal/nomic roles. The relationship between the kinds of property an object possesses and its specific behaviour in a possible world w is determined by the metaphysically contingent laws of w and so is itself contingent. Defenders of the neo-Humean view are reluctant to accept any *primitive* modal facts about the actual world (i.e., facts related to possibilities, necessities, etc.) and try to metaphysically analyse them in terms of the actual (and in some cases possible) fundamental *categorical* properties and the (nomic) relations holding between them.

In opposition to neo-Humeanism, non-Humeans *traditionally* defend an ontological framework based on the rejection of the intrinsic inactiveness characterising actual fundamental properties according to the

V. Livanios, *Science in Metaphysics*,
DOI 10.1007/978-3-319-41291-7_1

neo-Humean account. Non-Humeans believe in the existence of intrinsically *active* entities characterised essentially by primitive causal *powers*. According to the non-Humean worldview, at least some of the fundamental features of the world are *genuinely dispositional* in the sense that they are neither reducible nor supervenient on a base of fundamental categorical features. The transworld identity (and more generally, the *de re* modal representation) of these genuine dispositional properties is intimately related to the essential causal powers that determine the behaviour of objects possessing them in *each* possible world.

There are (at least conceptually) two *distinct* issues relevant to our purposes and related to the ongoing debate between neo-Humeans and non-Humeans. The first has to do with the broadly Humean claim that there are no *de re* necessary connections between wholly distinct existences. It is an issue related to the *de re* modality of the entities of the actual world; we cannot reach a verdict on that issue unless we have an answer to the question about the proper way of their *de re* modal representation. Especially in the context of metaphysics of natural fundamental properties, the relevant question concerns the way of the *de re* modal representation of properties.

The second issue concerns the truthmakers of a subset of modal truths of the actual world. I am primarily interested in the truths about fundamental interactions and temporal evolution of states of elementary physical systems,[1] and the question is: Are the actual fundamental (first-order) features *sufficient* truthmakers for these truths or not? This, in turn, is related, first, to the demarcation criterion between dispositional and categorical properties and, second, to the genuine actual existence of one type or both (or neither). For many philosophers the dispositional/categorical distinction and the nature of genuinely existing fundamental features comprise the two most crucial topics in the controversy between Humeans and non-Humeans.

The orthodox approach is to *exclusively* associate dispositional features with a specific way of a *de re* modal representation, while associating categorical features with another. As we shall see in Sect. 4.4, however, we have reasons to dissociate the two issues and adopt an alternative view

[1] I do not take sides here on the issue of whether those truths can be interpreted as *causal* or not.

according to which *all* kinds of fundamental features are *de re* modally represented in the *same* way. But before articulating this account, we should first comment on the dispositional/categorical distinction.

1.2 The Truthmaking Criterion for the Dispositional/Categorical Distinction

The remarks of the previous section might falsely create the impression that the majority of metaphysicians agree on the possible existence of two *clearly* distinct *kinds* of natural fundamental properties (the dispositional and the categorical ones) or two different *ontological* features (dispositionality and categoricality) which each fundamental property may instantiate. But this is far from being true. As far as the issue of a clear demarcation is concerned, it seems that there is a perspicuous *conceptual* distinction between dispositional and categorical properties. *Intuitively*, the former are related to *possible* manifestations of specific behaviours provided that appropriate triggering conditions will occur, while the latter seem to be *unconditionally* manifested properties that characterise objects as they 'actually' are. However, it has proved to be extremely difficult to turn this intuitive clarity into a theoretically adequate demarcation criterion.

According to some philosophers, the dispositional/categorical distinction is *not* ontological and concerns either the predicates we use (Mellor 2000, 767) or the ways we represent and characterise properties (see Mumford's earlier (1998) view, according to which properties can be *characterised* either dispositionally or structurally [categorically], relative to a particular causal role). Others (Martin (1997), Heil (2003) and Strawson (2008)) claim that dispositionality and categoricality are the self-same property differently considered, and that there are no ontological features grounding (or simply being) the dispositionality and categoricality of any property. In opposition to the above-mentioned authors, a number of metaphysicians hold that there is an *ontological* distinction between dispositional and categorical properties—a distinction which is not about predicates but rather about specific ontological features of properties in question. I, too, prefer to think of the dispositional/categorical distinction as an *ontological* issue mainly due to my qualms about Neutral Monism (which is the view according to which properties are by themselves neither dispositional nor categorical).[2] This places on me the burden of offering an appropriate ontological criterion of demarcation based on ontological features which might characterise fundamental properties and relations of the actual world.

Before presenting my preferred criterion, I should stress an important point. The well-known debate about the proper criterion of the dispositional/categorical distinction was largely taking place (and still takes place) at the *semantic* level. To that extent, one may consistently hold that the whole debate has *no* ontological consequences. For instance, even if a version of conditional analysis[3] proved to be problem-free, there can be no ontological reduction or elimination of dispositional properties in favour of categorical ones. This does not mean, however, that in our search for an ontological criterion for the dispositional/categorical distinction we ought not to take our cue from the attempts to find an adequate conceptual distinction between dispositional and categorical concepts. The issue is more subtle. On the one hand, it seems reasonable *not* to seek the proper ontological factor which may ground an ontological criterion among the ontological correlates of notions which proved to be insufficient to ground the conceptual distinction. On the other hand, even if a conceptual link proved to be promising for delineating the semantic/conceptual distinction, we must be careful to single out the *proper* ontological link corresponding to it. I suggest that this is the case with the famous relationship between dispositional ascriptions and counterfactuals. Given that we accept that there is a link between them,

[2] For a critique of Neutral Monism, see Sect. 2.2.2.

[3] According to the conditional analysis, the ascription of a dispositional predicate to an object *entails* the truth of (a) specific counterfactual(s) and vice versa. More precisely, the analysis proceeds in two steps: First, a dispositional predicate D is associated with specific stimulus C and manifestation M such that an object x has D iff x is disposed to M in C. Second, an analysis of the overt dispositional locution 'x is disposed to M in C' is carried out in terms of the counterfactual conditional 'If x were in C, x would M'.

there is extra work to be done for the identification of the proper ontological relation 'mirrored' in that link. One may, for instance, claim that there is a *direct* relation between having a dispositional property and a counterfactual *fact*, or, alternatively, that the relation is *indirect* because it is grounded in the *truthmaking* relation between dispositional properties and specific modal truths which, in turn, are expressed by appropriate counterfactuals.

A convenient strategy to find the proper ontological criterion of the dispositional/categorical distinction for fundamental properties consists in focusing on the ontological features of dispositionality, revealing the key issues associated with their possession and then defining categoricality mainly in contradistinction to them. To begin with, the possession of the ontological feature of dispositionality is related to the following two important issues. The first one has almost become a platitude; non-fundamental dispositional properties have *something* to do with the counterfactuals describing causal[4] truths which we all accept. I find nothing objectionable to the 'extension' of this common view to the case of fundamental properties, to the extent, of course, that the relevant *kind* of truths—perhaps non-causal in the *ordinary* sense—and the specific members of that kind are specified by mature and successful fundamental scientific theories. In spite of some views to the contrary, I think that the whole battery of counterexamples[5] to the conditional analysis only affects

[4] The set of specific modal truths associated with a fundamental dispositional property might include situations where, by definition, no 'ordinary' manifestation occurs. It is arguable, then, whether the truths related to those cases can be construed as *causal*. For a causal interpretation, see Mumford and Anjum (2011, 29–30).

[5] Martin (1994) argued that paradigmatically dispositional predicates do not entail the relevant counterfactuals. In his examples, Martin exploited the temporal interval between the appropriate triggering stimulus and the characteristic manifestation of a disposition to introduce hypothetical (external to the object having or lacking the disposition in question) *finks* or *counter-finks* which eliminate the disposition an object possesses or 'generate' the disposition an object lacks, respectively. Bird (1998) introduced *antidotes*, that is, factors which by definition leave the alleged causal basis of the disposition intact but break the causal chain leading to the manifestation M of the disposition, so that M does not occur. There are also *counter-antidotes* or *mimics*; they are external-to-the-causal-basis factors which 'generate' a manifestation of a 'disposition' an object actually lacks when the appropriate stimulus has occurred. The literature on the conditional analysis of dispositions is huge; see, for instance, Carnap (1936; 1937; 1956), Mellor (1974; 1982), Prior (1982), Lewis (1997), Mumford (1998), Malzkorn (2000), Gundersen (2002), Bird (2004), Cross (2005), Choi (2005; 2008; 2012; 2013), Manley and Wasserman (2007; 2008), Clarke (2008), Handfield (2008), Everett (2009), Ashwell (2010), Manley (2012) and Contessa (2013).

its prospects, but does not refute the existence of *some* kind of ontological link lurking in the shadow of the relationship between dispositional ascriptions and counterfactuals.[6] Following Bostock (2008) and Jacobs (2011),[7] I suggest that the relation between the instantiation of a fundamental dispositional property and the truth of specific counterfactual(s) is a *truthmaking* one.[8] A second issue is related to the broader Humean/anti-Humean debate. Fundamental dispositional properties constitute the ground for the so-called active worldview according to which properties confer *by themselves* powers on their bearers.

Combining the above 'truisms' about dispositional properties, we reach the following criterion of dispositionality:

@-criterion of Dispositionality The first-order state of affairs of an object instantiating a fundamental dispositional property is *by itself* (part of)[9] a minimal[10] truthmaker for specific (perhaps, world-relative) modal truths (expressed by specific non-trivial counterfactuals) which concern the bestowal of specific (perhaps, world-relative) powers[11] on the object.

The @-criterion of dispositionality allows the instantiation of a fundamental dispositional property to be (part of) the minimal truthmaker for more than one modal truths (perhaps an infinite number of them). This could cause no trouble for its plausibility since it is not supposed that the truthmaking relation is necessarily injective in the first place.

---

[6] A recent example of a philosopher who refutes the existence of an ontological link is Ingthorsson (2013). He argues that we have no compelling reason to think that the possible behaviour of the bearers (expressed by counterfactuals) has anything to do with the essence (and plausibly with the criterion of demarcation) of dispositional properties. For Ingthorsson, the specification of the relevant counterfactuals is of purely epistemological import.

[7] In fact, Jacobs defends a kind of identity view about properties and so his suggested definition explains in part the conceptual distinction between dispositionality and categoricality which are, in his view, identical.

[8] See, however, Melia's (2005) claim that it is not necessary to think of truthmaking as a relation.

[9] The qualification is necessary because, as Jacobs clearly explains, most counterfactuals involve many properties, which only if taken collectively will be eligible to be *the* truthmaker.

[10] If e is a minimal truthmaker for a truthbearer B, then we cannot subtract anything from e with the remainder still being a truthmaker for B. For a discussion, see Armstrong (2004, 19).

[11] In our world, these specific (causal?) powers are determined by the findings of our fundamental, mature and successful physical theories.

Moreover, this is a desired feature according to some believable (albeit not unanimously accepted) views about fundamental properties, according to which these are *determinable* quantities having an infinite number of determinates falling under them and entering in a variety of laws (consider, for instance, the property of electric charge q which may take infinite discrete values q_i and appears in Coulomb's law, the Biot–Savart law, etc.). Both of these facts indicate (for the advocates of these views) that fundamental properties can be *multi-track* dispositional properties associated with a large (most probably, infinite) number of counterfactuals expressing modal truths. Notice also that @-criterion does not require of the state of affairs of an object instantiating a fundamental dispositional property to make true the *same specific* truths in all worlds in which it obtains.[12] (I take it that the obtaining of the state of affairs in a possible world does not presuppose the transworld existence of either the object or the dispositional property. So the previous modal assertion can be interpreted as claiming that a counterpart of the object instantiates a counterpart of the property.) It does, however, require of the state of affairs of an object instantiating a dispositional property to be *by itself* (part of) a minimal truthmaker for some (at least world-relative) specific modal truths.

In contradistinction to the above criterion of dispositionality, I suggest the following criterion of categoricality:

@-criterion of Categoricality The first-order state of affairs of an object instantiating a fundamental categorical property is *not* by itself (part of) a minimal truthmaker for specific (perhaps, world-relative) modal truths (expressed by specific non-trivial counterfactuals) which concern the bestowal of specific (perhaps, world-relative) powers on the object. In order to be (part of) a minimal truthmaker for the aforementioned truths, it must be supplemented with a *nomic* fact relating the property in question and other properties and/or relations.

@-criteria are clearly ontological, since they identify ontological factors (states of affairs) the different truthmaking role of which grounds the dispositional/categorical distinction. Furthermore, in my view, they

[12] Thus, it leaves room for unorthodox versions of dispositionalism like the one defended by Hendry and Rowbottom (2009).

'capture' the core elements of the distinction as presently refined through philosophical analysis.[13] Before I conclude this section, I have two final remarks to make: First, some philosophers (see, for instance, Dodd (2002), Hornsby (2005), Lewis (2001), Melia (2005)) who believe in *truthmaking* do not also believe in *truthmakers*, that is, they do not hold the view that truth is grounded in *entities*. They rather think that the ground of truth resides *in how entities are*. Furthermore, even among advocates of truthmakers, there is no consensus on which entities *can* be truthmakers. Some opt for states of affairs or facts, while others postulate non-transferable tropes for the truthmaker-role. As far as I can see, one does not have to take sides on this debate in order to embrace @-criteria. For though stated in terms of states of affairs, @-criteria can be easily expressed in terms of tropes or even without truthmakers at all. In the first case, the minimal truthmaker would be a non-transferable trope of the fundamental property, while, in the second, the ground of the specific modal truths involved in the criteria would reside in how the fundamental property really is. Second, there is an ongoing debate about the very definition of truthmakers. According to a first account, an entity e is a truthmaker for the proposition <P>[14] iff e exists and the proposition <e exists> *entails* <P>. An essence-based view suggests instead that e is a truthmaker for <P> iff it is part of the essence of <P> that <P> is true if e exists. Though I am sympathetic to Rodriguez-Pereyra's 'in virtue of' view (according to which e is a truthmaker for <P> iff <P> is true in virtue of e—the notion 'true in virtue of' is regarded as primitive), I won't discuss this issue further, since I do not think that anything of relevance for my present purposes hinges on it (for a brief discussion, see Rodriguez-Pereyra 2006).

[13] A note of caution: The fact that dispositional properties are constituents of states of affairs which are (parts of) minimal truthmakers for (perhaps) infinite truths does not entail the infinite ontological complexity of properties themselves.

[14] Here it is assumed that the truthbearers are propositions, but nothing (as far as the suggested definitions are concerned) depends on that assumption.

1.3 Arguing Against Dispositional Monism from the Actual Existence of Fundamental Categorical Features

In the previous section I presented my preferred account of the dispositional/categorical distinction. This is the necessary background material in order to proceed now to the examination of the various theories of the nature of fundamental properties. My ultimate aim is to defend (in Chaps. 3, 4, 5 and 6) a sui generis kind of Categorical Monism, according to which there actually exist only *contingently categorical* fundamental features. But before that, I have to examine the other metaphysical accounts of the nature of properties and lay down my own reasons to reject them. To this end, I start in this chapter by examining another kind of property-monism; that is, Dispositional Monism.

Categorical Monism (in its 'traditional' form) was orthodoxy in analytic circles, but since the resurrection of powers (mainly through the seminal works of Harre (1970), Harre and Madden (1975) and Shoemaker (1980)), realism about ontologically genuine dispositional properties has gained growing popularity and, with it, the plausibility of Dispositional Monism was significantly raised. By definition, Dispositional Monism is the metaphysical view according to which:

All fundamental natural properties are purely dispositional.

Contrary to most of the literature on the subject, I do not intend in this chapter to discuss the challenges Dispositional Monism had to meet in order to prove its viability against Categorical Monism.[15] I would rather present what I think is the main reason for rejecting Dispositional Monism.

It is certainly true that the most straightforward way to refute any property-monism of a certain kind is to prove the actual existence of properties of a *different* kind. Yet, this is not always an easy task; that is

[15] For a thorough exposition of relevant arguments, I definitely suggest Alexander Bird's (2007) excellent work, where one can find not only responses to a large number of objections but also reasons to prefer Dispositional Monism against Categorical Monism.

why philosophers often follow the implicit course and either offer various arguments against the thesis or argue in favour of the superiority of its rivals. Nevertheless, I strongly believe that, adopting the straightforward way, we may adequately argue against the thesis of Dispositional Monism. In particular, given my preferred ontological criterion of the dispositional/categorical distinction, I'll now present my arguments for the existence of actual fundamental *purely categorical* features. More precisely, I'll argue that, according to the most successful theory of spacetime that we currently have, General Relativity (GR), the spatiotemporal relations (as encoded in the metric field) are categorical features of our world.

1.3.1 Spatiotemporal Relations as Fundamental Categorical Features

1.3.1.1 Introduction

Spatiotemporal relations are admitted as genuinely existing both by substantivalists and by relationists about spacetime. Substantivalists think of them as holding primarily between spacetime points and only derivatively between material objects or events, while relationists insist that they are relations between exclusively material objects.

Given the current state of the debate about the nature of fundamental physical properties, it seems odd that I need extra arguments to defend the categoricality of spatiotemporal relations. For, this is not a view held by categorical monists only; some philosophers who believe in genuine dispositional properties also claim that the existence of non-dispositional features is indispensable in order for all causal laws to be operative. For instance, both Molnar (2003) and Ellis (2001; 2005) highlight the peculiar role that spatiotemporal properties and relations play in non-Humean metaphysics. Molnar (2003) argues that most of the dispositional properties of objects are location-sensitive; which determinate of the determinable (of the property) is manifested on a particular occasion depends on the particular value taken by the variable location. So, although for him positional properties are categorical, they have nonetheless an ineliminable role in the manifestation of dispositional properties. It seems that Molnar admits a form of (ontological) dependence of dispositional properties on categorical ones, but he is not clear about it.

Brian Ellis attempts to be more specific about the role of spatiotemporal properties. He suggests that a property may have a causal role (in the sense that it is a relevant causal factor) without being either a causal power or ultimately reducible to causal powers (2005, 470). A kind of such properties (which include spatiotemporal properties and relations) is the *dimensions* of causal powers; that is, essential properties of powers which determine their location, distribution, magnitude, number and mathematical (scalar, vector, tensor, etc.) nature. According to Ellis, these properties are categorical. They do not confer of necessity any powers on things. For instance, although the spatiotemporal separation of any two objects is a real objective relation holding between them, it does not essentially require them to behave in any particular manner. Nevertheless, dimensions will need to be specified if we are to adequately define what the powers of things are (*ibid.*, 471).

Contrary to the above 'orthodox' view, some authors have suggested (or can be interpreted as suggesting) that spatiotemporal features have a dispositional character. For instance, Mellor's (1974) remark that possessing geometrical properties (such as 'is triangular') entails subjunctive conditionals can be interpreted as a suggestion that these prima facie categorical properties are also dispositional.[16] Given that these properties exist in virtue of the spatial relations holding between the parts of the objects possessing them, they can be broadly construed as spatiotemporal properties, though non-fundamental ones. Hence, one may plausibly argue that Mellor's point vindicates the view that (at least some) non-fundamental spatiotemporal properties are dispositional. Since, however, the metaphysical accounts of the nature of properties concern (in my view) the *fundamental* ones, Mellor's points are irrelevant to the present discussion.[17]

---

[16] According to one popular criterion of dispositionality, the entailment of subjunctive conditionals is a distinctive feature of dispositional properties.

[17] Alexander Bird (2007, 160 & 167–8) also thinks that the whole discussion should concentrate on *fundamental* metrical relations and not on familiar properties such as shape which are geometrical (structural). He thinks that even if structural properties fail to satisfy any criterion of disposi-

Recently, Alexander Bird (a prominent defender of Dispositional Monism) has also suggested the dispositionality of spatiotemporal relations. Bird (2007; 2009) aims to defend Dispositional Monism against the 'orthodox' view by providing a dispositional account of spatiotemporal relations such as the spatial distance between two material objects or the temporal distance between two events. Other philosophers such as Bartels (Bartels 1996; 2013) and Esfeld and Sachse (2011) have also argued for the dispositional character of spatiotemporal relations insisting on their *causal* efficacy.[18] The arguments of Bird, Bartels and Esfeld are extremely relevant to the present discussion for two reasons. The first one is that they refer to the most plausible candidates for fundamental relations in our world and not to some derivative, non-fundamental actual features. The second is that they are motivated by the findings of the best theory we currently have for spacetime, Einstein's GR. Hence, they have the credentials to be the basis for a proper argumentation in the context of a science-informed metaphysics.

In what follows, though I'll focus on Bird's explicit endeavour to endow fundamental spatiotemporal relations with dispositional essence, I'll have the opportunity to critically discuss the arguments advanced by Bartels and Esfeld. Bird's argument is cashed out in terms of a modal criterion of dispositionality based on subjunctive conditionals. For Bird, in all possible worlds, any object possessing a dispositional property P is disposed to yield a particular manifestation M in response to a characteristic stimulus S. Arguably, a conditional analysis of the foregoing disposition can be offered by appealing to subjunctive conditionals. This is not the place to review the difficulties besetting this analysis, largely due to the presence of finks and antidotes. Nevertheless, Bird introduces the following test based on conditional analysis in order to decide whether spatiotemporal relations are dispositional or not:

tionality, they would not provide counterexamples to the monist's claim which concerns *only* fundamental properties and relations.

[18] Bartels (1996) talks about spatiotemporal metrical relations as providing *causal capacity* to spatiotemporal points. If that were true, it would allow us to ascribe a causal–dispositional essence to spatiotemporal relations. (Of course, since Bartels' main issue in that paper is the individuation of spacetime *points*, he is not trying to convince us that spatiotemporal *relations* themselves have dispositional essence.)

> "P" denotes an essentially dispositional sparse property iff for all x, and for some stimulus S and manifestation M, x is P entails were x to be S, then x would be M. (Bird 2009, 217)

The above test fails to provide a necessary condition for a property being dispositional due to the possible presence of finks and antidotes. However, as Bird (2007) carefully argues, neither finks (2007, 60) nor antidotes (2007, 62) could occur at the fundamental level.[19] Hence, insofar as he only cares about fundamental spatiotemporal relations, finks and antidotes cannot affect the adequacy of his test. Moreover, the suggested test also fails to provide a sufficient condition for a sparse property being essentially dispositional due to several reasons that I am not going to discuss here. Nevertheless, the lesson which Bird draws is that we must be careful in employing the test rather than reject it. In the following discussion, I am not going to express any qualms about the above contentions. Rather, I take them for granted and examine whether spatiotemporal relations may pass the suggested test and be construed (according to Bird's test) as possessing a dispositional essence. My aim is to show that the state of affairs of spacetime instantiating a metric structure cannot by itself be related to a 'proper' (according to Bird's test) counterfactual in the actual world, and so, can neither be by itself a truthmaker for a 'proper' specific actual truth in order for the spatiotemporal relations to be dispositional in our world.[20]

1.3.1.2 Spatiotemporal Relations and Subjunctive Conditionals in Pre-GR Theories

I begin the discussion with a brief overview of theories prior to GR and seeing whether, according to them, spatiotemporal relations entail any non-trivial subjunctive conditionals. The main feature of pre-GR

[19] Bird (2007, 63) thinks that the case for fundamental antidote-free dispositions is less clear than the case for fink-free ones. Nevertheless, as he notes, the development of physical science shows that the prospects for antidote-free fundamental properties are promising.

[20] For those who reject the existence of a spacetime manifold of points instantiating the metric structure, it is the metric field itself which might be the truthmaker in question.

theories is that spacetime is fixed independently of any processes and events occurring within it. According to the substantivalist conception of spacetime, spatiotemporal relations hold between spacetime points and only derivatively between material objects (which occupy the points). Hence, given the fixed spatiotemporal background, spatiotemporal relations cannot be recipients of change and (following the action–reaction principle[21]) cannot be agents of change either. In the pre-GR context, therefore, spatiotemporal relations, understood in substantivalist terms, cannot have a dispositional essence because they cannot bring about (or, at least, probabilify) any effects on matter.

Things are different, however, from the relationist perspective. Relationists claim that, since spatiotemporal relations hold between material objects, they can change due to physical interactions between bodies. Indeed, Newton's second law demonstrates how spatiotemporal relations between objects change given that a certain interaction takes place. Although it is reasonable, from this perspective, to think of spatiotemporal relations as passive dispositions, we cannot plausibly claim that they also are active dispositions (agents of change). Consider, for instance, a possible manifestation of Coulomb's law. It seems unintuitive to regard the spatial displacement r as a disposition the manifestation of which is the force acting between the charges. Mumford (2004, 188) challenges this common view suggesting that, in laws such as the this, we may characterise dispositionally not only charges but also the spatial relation between them. Consider, for instance, two specific instantiations of point charges q_1 and q_2 separated by a spatial distance r. The suggested account is the following:

> [(Two specific instantiations of) q_1 and q_2 are terms of r] entails [if (these instantiations of) q_1 and q_2 were separated by r, a force would be acting between them with magnitude $F=kq_1q_2/r^2$][22]

[21] Einstein endorses the action–reaction principle as a scientific principle and insists that it is contrary to the mode of scientific thinking to conceive of a thing which acts itself, but which cannot be acted on. Leibniz is one philosopher who thinks that conforming to this principle is a defining feature of substances.

[22] Strictly speaking, the entailment holds between the *propositions* expressing the content of sentences in brackets.

Bird (2007, 161) criticises this suggestion arguing that Mumford's move can also be applied to *any* law sensitive to spatial distances (for instance, to Newton's law of gravitation with point masses m_1 and m_2 separated by r), and, therefore, yields a *multi-track* disposition (one with a more than one kind of manifestation). Bird thinks that this poses a problem for the dispositional monist,[23] because it is not reasonable to expect that the conjunction of the distinct dispositional essences (corresponding to each kind of manifestation) is equivalent to one dispositional essence related to r. This remark, however, does not settle the issue because there are alternative courses for the monist to take. For instance, she may argue that in the case of r we don't have a *single* disposition, but many; m_i 'having' r, q_i 'having' r, and so on. Alternatively, she may follow Martin (1997) and argue that m_i, q_i, and so on, are distinct *dispositional partners* of r, hence, one and the same disposition r can have different kinds of manifestation with different kinds of reciprocal partners. Bird himself seems to prefer another course in order to maintain Mumford's insight. He thinks that one kind of manifestation is *privileged* among the others. It is the one that is related to the law of gravitation. One may wonder what grounds his choice. It is here that he abandons the pre-GR context and focuses on the general relativistic framework. In his own words,

> given the general theory of relativity, it is natural to see gravitational force as participating in the essence of spatial properties and relations. (2007, 162)

Bird's choice is clearly justified in the context of science-informed metaphysics because Einstein's masterpiece is the best spacetime theory we currently have.

1.3.1.3 Bird's Argument for the Dispositional Essence of Spatiotemporal Relations

Bird applies the test previously presented in order to show that fundamental spatiotemporal relations pose no threat to Dispositional Monism.

[23] The problem arises for a dispositional monist who thinks that spatiotemporal relations are fundamental. In the opposite case there is no problem, because the causal inefficiency of derivative relations does not (arguably) pose any threat to the monist.

His argument, however, relies on the validity of Einstein's equations[24] which refer directly to the metric field and not to the metrical (spatiotemporal) relations. It seems, therefore, that the argument to be presently considered is most naturally understood as one concerning the putative dispositional essence of the metric field. Since, however, the metric field 'incorporates' in a sense the totality of the metric spatiotemporal relations, in what follows I assume that Bird's argument is (at least implicitly) related to the dispositional essence of spatiotemporal relations.

In order to reconstruct the argument, let us first recall that, according to Bird's test, we may ascribe a dispositional essence to a property or relation P iff we can find *appropriate* stimulus and manifestations conditions (S and M, respectively) such that, for all x, Px entails (ceteris paribus) the counterfactual 'had S occurred, M would have occurred'. Now, in the case at hand, x refers to spacetime (or the totality of spacetime points, if you like), while P is the metric structure. Bird argues that the appropriate stimulus S is a variation of the total matter-distribution of the world, while the proper manifestation M is the associated variation of the metric structure of spacetime. So here is his entailment test applied to the case of spatiotemporal relations:

> Metric structure is essentially dispositional iff the proposition that the state of affairs of spacetime instantiating that structure obtains *entails* the counterfactual that were the total matter-distribution of the world to undergo a variation, the metric structure of spacetime would undergo a variation as well.[25]

Bird suggests that what makes the entailment relation true (and, in his view, enables us to ascribe a [passive] dispositional essence to the metric) is the hypothesis about the *background independence* of any true physical theory. According to that hypothesis (endorsed by some eminent quantum gravity theorists such as Baez (2001) and Rovelli (2004)), a

---

[24] Einstein's equations are six independent differential equations having the form $G_{\alpha\beta} = 8\pi \cdot T_{\alpha\beta}$ where $T_{\alpha\beta}$ is the stress-energy tensor (related to the distribution of mass–energy in the universe) and $G_{\alpha\beta}$ is the Einstein's tensor (related to the curvature of spacetime and constructed from the metric tensor $g_{\alpha\beta}$ and its derivatives).

[25] Once again, one might say that it is the existence of the metric field itself which entails the relevant counterfactual.

true physical theory should be characterised by a complete lack of *non-dynamical* entities or structures. The only non-dynamical entity which pre-GR physical theories were admitting is spacetime. Hence, it comes as no surprise that the definition of the concept *dynamical* is intimately related to spacetime structures (although it can be defined more generally). A dynamical spacetime structure is one which is not fixed independently of the processes and events occurring within spacetime. Or, in rough (albeit misleading, strictly speaking) terms, it is one which 'affects' the material content of spacetime and is 'affected' by it in turn (Friedman 1983, 64). Given all that, one may apply Bird's test to the metric structure and arguably ascribe to it a *passive* dispositional essence. Finally, it is important to note that the truth of the counterfactual 'were the total matter-distribution of the world to undergo a variation, the metric structure of spacetime would undergo a variation as well' ensures only that, under appropriate stimuli, the structure of spacetime is changed. It does not, however, guarantee that spacetime structure itself is responsible for changes. Hence, in order to ascribe an *active* dispositional essence to the metric, we need to appeal to the *action–reaction principle* which tells us that since spacetime and its properties may be the recipients of change they may also be causes of it.

1.3.1.4 An Appraisal of the Argument

Bird's argument rests, inter alia, on scientific authority. A prima facie objection to it, however, is that it rests exclusively on *speculations* of currently developing physical theories. This would certainly be the case were the argument to rest solely on the radical interpretation of background independence according to which *future* physical theories should eliminate spacetime from the inventory of *fundamental* entities of our world. (I take it for granted that Dispositional Monism—just like the other accounts examined in this work—is a metaphysical view concerning the fundamental level of entities.) However, this appeal to future science (if it were exclusive) would indirectly undermine several arguments from scientific authority that dispositional realists have already used to argue for the dispositional character of the other fundamental properties. For, if current physical science (which undeniably posits spatiotemporal relations) is regarded as 'unreliable' for the metaphysical analysis of spatiotemporal relations, then on what grounds do we rely on its findings in the case of other fundamental physical properties such as mass, charge, and so on? Fortunately for Bird's case, background independence is a central feature of the most successful spacetime theory that we *currently* have, GR (and this, I think, is the main reason why some theoretical physicists 'project' this feature, as a basic principle, to future theories). According to GR's less radical interpretation of background independence, spacetime exists but is not the *fixed* background that physicists used to think; it is rather a *dynamical* entity (or dynamical structure). The bad news for Bird's case is that this modest interpretation of background independence cannot offer any conclusive reasons to believe that spatiotemporal relations are indeed dispositional. Let us see why.

I take it as a truism about natural properties/relations that their physical existence may be related to a wide range of different counterfactuals, though only some of them (or just one) reflect their alleged dispositional essence. As Bird (2007, 158) himself notes, the search for the 'proper' counterfactual may be guided by the following test: The stimulus and manifestation present in the counterfactual should reflect the role of the property/relation in scientific explanations. Some passages in Bird's 2007 publication suggest that he prefers a *causal* interpretation of the counterfactual describing the dispositional essence of spatiotemporal relations. For example, in page 168, he says:

> On a classical substantivalist conception of space, spatial separation is a relationship between points in an unchanging spatial background and thus incapable of acting as a *cause* and *so also incapable of having a dispositional essence* (My italics).

A few lines below:

> In the light of action-reaction principle it is the fact that in classical physics space is a mere background that prevents us from being able to regard it as having a *causal role* and *so prevents us from seeing it as having a dispositional essence* (My italics).

However, it might be that the explanatory role of a fundamental natural property or relation has nothing to do with causality as such but is related instead to its broader *nomic* role. Returning to Bird's 2007 work, we find another passage (2007, 160), where in the course of commenting on a suggested counterfactual for the case of the structural property of triangularity, he notes that that conditional fails to get at the heart of the problem because

> it is difficult to see that anything like a causal *or nomic role* is being assigned to triangularity (My italics).

It is not clear, then, which of the two interpretations Bird has in mind. In what follows, I'll begin with assuming that he aims to interpret his suggested counterfactual causally. That is, I'll presuppose that the truth of the counterfactual amounts to the fact that the variation of the total matter-distribution of the world *causes* an associated variation of the metric structure of spacetime. After that, I'll also discuss the possibility of a *nomic* interpretation of the suggested counterfactual.

'Affection' of Matter on Spacetime

In the context of GR matter 'affects' spacetime in a precise manner described by Einstein's equations. So the modal behaviour of matter (and also of spacetime), which is vaguely asserted through the counterfactual of Sect. 1.3.1.3, is precisely described by the basic equations of GR. Assuming a causal interpretation of Bird's counterfactual compels us to read Einstein's equations as saying that the curvature of space–time depends *causally* on the distribution of matter fields throughout it. The causal interpretation of the influence of the distribution of matter on spacetime can however be challenged in the context of a variety of available accounts of causation. Consider first the regularity accounts of causation which define the latter in terms of regularities of the *actual* world. According to these accounts, the causal interpretation of Einstein's equations implies that any variation of the matter-distribution in the actual world must *change* the space–time structure of *our* world. Notice, however, that we talk about the *distribution* of matter, and the very notion of distribution *presupposes* an already existing metric structure. As Sklar (1976, 75) rightly notes:

> the stress-energy tensor takes into account the distribution of mass-energy in the world utilizing the *metric* features of this distribution.

The *dynamical* character of metric structure in GR implies that the metric is *not* an absolute object postulated by GR. But, crucially, all it means for an object to be non-absolute in the technical sense which is relevant here is that it is not the 'same' (up to diffeomorphism-equivalence) in every dynamically possible model of GR (Friedman 1983, 64). Given that most plausibly each class of diffeomorphically related models of GR[26] corresponds to a dynamically possible world described by the theory, the dynamical character of space–time allows only the *transworld* variation of spacetime structure in such a way that Einstein's equations hold. According to that approach, we look upon the equations as giving a law-like consistency constraint on the joint features (space–time structure and mass–energy distribution) of any (physically) possible world. Here is Sklar again:

> We describe a world by picking a coordinate system, expressing a g-function [metric] in terms of that coordinate system which characterizes the space-time of the world, and then picking a stress-energy tensor function to describe the distribution of the non-gravitational mass-energy and forces throughout the world. If the proper function of the g-function we have chosen is equal to the stress-energy function, then the world is genuinely possible according to General relativity. Otherwise, what we have described is not a possible world at all. (Sklar 1976, 215)

Bartels (2013, 2005), while commenting on an earlier (Livanios (2008)) presentation of my critical thoughts concerning Bird's argument, objects

---

[26] A model of GR is < M, Oi>, where M is a differentiable manifold and Oi the geometrical objects defined on it (i.e., the metric tensor g and the stress–energy–momentum tensor T). d* is the carry-along map induced by the diffeomorphism d. A crucial feature of GR, general covariance, ensures that given a model A = <M, g> of GR and a diffeomorphism d: M → M, A′ = <M, d*g> is also a model.

to the aforementioned interpretation on the grounds that Einstein's equations *can be used* to describe local variations in matter-distributions (e.g., gravitational collapse of a star) of a world like ours which lead to drastic changes in the local spatiotemporal structure of the *same* world. Yet, in the four-dimensional framework there is no change understood in this way; the whole set of events describing, for instance, the gravitational collapse of a star is compatible with a metric structure (plus boundary conditions) associated with a solution of Einstein's equations for *all* spacetime. Such a solution describes a specific world and no alteration of its *four-dimensional* structure occurs *intraworldly* (see also Smart (2013)).

Given the previous remarks concerning the prospects of a regularity-account-of-causation-based analysis of the putative causal impact of matter on spacetime, one may plausibly think that the appropriate context for such a causal interpretation is within a *counterfactual* account of causation. As is well known, the latter defines causation in terms of counterfactual dependence of the effect on the cause. To be more precise, according to a counterfactual account, when C causes E, the counterfactual conditional 'If C had not occurred, E would not have occurred' must be true. In the case at hand, the following counterfactual must be true:

> (CDMM) If variation of matter-distribution had not occurred, variation of metric structure would not have occurred.

According to the standard possible world semantics, CDMM (Counterfactual Dependence of the Metric on the Matter-distribution) will be true if there is a possible world in which there is no variation of both matter-distribution and metric structure that is *closer* to the actual world than any possible world in which metric is different though matter-distribution remains the same. In the context of GR, however, a stress–energy tensor (describing the matter-distribution in a world) is, in general, compatible with many different metric structures. For, the Einstein equations by themselves do not specify a unique metric associated with a given matter-distribution; we need also the imposition of specific *boundary* conditions. Crucially, all possible worlds with the same matter-distribution but different metrics share by definition Einstein's law (they are all models of GR), and, furthermore, there is no reason to think that they do not share the other laws of nature as well. Given then that the notion of world-similarity needed for the evaluation of the truth-value of the counterfactual in question is most commonly based on laws-sharing between possible worlds, I see no obvious reason to claim that worlds with different metric structure but the same matter-distribution with the actual one are *less* similar to the actual world than worlds where both metric and matter-distribution differ from their actual 'values'. Hence, I conclude that CDMM is most probably not true. Despite its initial appeal, the counterfactual account cannot be the proper context for a causal interpretation of the influence of matter on spacetime.[27]

Let me finally turn to Dowe-Salmon's conserved quantity account (CQAC) to see whether it can successfully accommodate the alleged causal impact of matter on spacetime. According to CQAC, causation is conceived in terms of causal processes (worldlines of material bodies) and causal interactions (intersections of worldlines involving an alteration of the value of a conserved quantity for at least one incoming and one outgoing worldline). A crucial feature of CQAC is that any change of a relevant quantity taking place during a causal interaction is governed by conservation laws among which (and perhaps the most important one) is conservation of energy. Yet, as various authors (physicists and philosophers alike[28]) have pointed out, the existence of a non-gravitational energy conservation law in a given spacetime region depends (according to GR) on specific symmetries which a *general* model of the theory does *not* possess. Indeed, not even models purporting to describe the actual world possess the desired features; those that do, represent highly idealised situa-

[27] Curiel (2015 manuscript) claims that physicists evaluate counterfactuals in the context of GR by selecting similarity-measures among possible worlds via *ad hoc* fixing of a comparison class of spacetimes (for instance, they require that the spacetimes in the class have a fixed topology or satisfy certain symmetries). For him, however, though this method is often justified by strong physical reasons, it remains *ad hoc* and, so, inadequate from the philosophical perspective. Curiel thinks that there is an inherent difficulty in giving a *general* account for the evaluation of counterfactuals in GR. He claims that the most natural way for such evaluation is to treat counterfactual changes as variations in *initial conditions* and then use the machinery of GR's initial-value formulation (for an exposition of the latter, see Wald (1984, Ch.10)). Since, however, there is no absolute background structure for a comparison of properties/structures of different solutions of Einstein's equations, one has a difficulty to make sense of the notion of 'different "values" of the same property/structure in different models of GR'. Curiel's arguments, if sound, provide a strong reason not to embrace the counterfactual account of causation in the context under discussion.

[28] Here is a short list: Anderson (1967, Ch.13), Misner et al. (1973, Ch.20), Hawking and Ellis (1973, 61–62), Wald (1984, 84–85 and 286–295), Curiel (2000), Hoefer (2000), Lam (2011).

tions that at best approximate the material structure of our world. But if there is no non-gravitational energy conservation law for any spacetime region, then the whole set of notions grounding CQAC (i.e., energy as a conserved quantity, possession of energy, exchange of energy) becomes problematic in the general relativistic context. It might be thought that, by taking into account the exchange of energy between non-gravitational and gravitational fields, we might finally find a genuine conservation law for the *total* energy which (arguably) supports the basic tenets of CQAC. That is a forlorn hope however; for there is no way within the context of GR (i.e., in a general model of the theory) to define unambiguously the gravitational energy content in any given spacetime region. I conclude that the prospects of implementing CQAC in the case of the influence of matter on spacetime are indeed very poor.[29]

Before discussing the nomic interpretation, we should point out that the only thing which (allegedly) can be proved given that matter *causes* changes in spacetime metric structure is that the latter is a recipient of change and consequently, according to Bird's test, does have a *passive* dispositional essence. Yet, as I have previously remarked, in order to ascribe an *active* dispositional essence to the metric, we need to appeal to the *action–reaction principle* which tells us that since spacetime and its properties may be the recipients of change they may also be causes of it. In the sequel, however, I'll argue that in this case the upshot of the implementation of the action–reaction principle is most possibly not true.

'Affection' of Spacetime on Matter

The crucial point behind the denial of a causal 'affection' of spacetime on matter is that it is common in physical theories to relate causation to the action of various *forces*. Given that, the spatiotemporal structure

[29] Here I do not discuss the Dispositionalist Account of Causation (DAC). In rough terms, according to DAC, every instance of causation consists in the manifestation of a genuinely dispositional property (or the mutual manifestations of reciprocal dispositional partners). It is in the very (dispositional) nature of properties to causally produce (or tend to produce) the instantiation of other properties (for a recent defence, see Mumford and Anjum (2011)). It is obvious, however, that, for the present purposes, one cannot appeal to DAC since it presupposes what she aims to prove, that is, the dispositional nature of the metric.

(better, the metric-compatible affine structure of spacetime) cannot causally affect matter because all it can do is to determine the purely inertial motion of material bodies, in which no forces operate at all. In the context of Special Relativity (SR), the affine geodesics of spacetime project upon linear paths in 3-space, and so there is no need to postulate any kind of force to explain the motion of bodies due to the 'influence' of spacetime. In GR, due to the variable non-zero curvature of spacetime, affine geodesics do not project upon linear paths of space; hence, prima facie, we need to postulate gravitational forces in order to explain why bodies which are free of (non-gravitational) forces deviate from the geodesic paths of 3-space. However, one of Einstein's key points in GR is that, even in this case, there is no need for the above postulation, because bodies move along an inertial path in curved *spacetime*. Hence, in my view, it is clear that in a four-dimensional ontological framework spacetime does not causally affect matter. All the geometry of spacetime does is to determine which paths are available to bodies when moving inertially (geodesics); it does not *force* bodies to move in a certain way. Inertial motion, therefore, is explained geometrically, not causally. One may even follow Nerlich (2010) and interpret the geometrical character of this explanation as a special case of explanation by citing *identities*. According to Nerlich, the deviation of geodesics *is* the curvature of spacetime, while the latter does not cause but *is* the acceleration of deviating space and time trajectories, some among which happen to be trajectories of material particles.

Bartels (2013, 2005) thinks that one cannot argue against the *causal* influence of spacetime on matter on the basis that the former applies no *force* on the latter. For him such a line of thought presupposes what GR allegedly denies; namely, that, in the context of classical physics, the notion of cause is *necessarily* related to the concept of force. Bartels thinks that one of the innovations of GR is precisely that, contrary to pre-GR theories, it introduces a notion of cause *independent* of the notion of force. This claim, however, seems to me to be unwarranted. Currently, there are three interpretations of what, according to GR, the influence of spacetime on matter amounts to: Some physicists and philosophers think of GR as 'geometrising' away the gravitational force, while others think of it as showing that all spatiotemporal phenomena are expressions of the gravitational field. Finally, there is a third view according to which, in the general relativistic context, there is a conceptual identification of the gravitational field with the proper spatiotemporal geometrical structure.[30] In neither of these interpretations we have an indication that GR *dissociates* the concept of the cause from forces. Besides, Bartels himself appeals to the 'anachronistic' concept of force by associating the putative causal influence of spacetime with the action of gravitational tidal *forces*.

Is there anything more that we can say about the phenomenon of tidal forces itself? In my view, if we want to keep calling them 'forces', we have to make clear that, in opposition to other forces, they have nothing to do with causation. Bartels seems to think that tidal forces have a causal influence because they bring about a spatial *deformation* of extended material bodies. But this deformation can be explained geometrically by appealing to the relative acceleration[31] of the *force-free* motion of the parts of bodies following the affine geodesics of spacetime. The affine structure explains the deviation of geodesics which intersect a space-like cross section of any extended material body, and this deviation explains in turn the alteration of elastic electromagnetic forces which gives rise to the deformation. The spatiotemporal structure, however, does not cause the forces; as Nerlich (2010) stresses, the causal story in that case is exhausted in the role of electromagnetic forces and has nothing to do with the influence of spacetime on matter. Besides, as Bartels himself acknowledges (*ibid*., 2006), tidal 'forces' exist (in the geometrical sense of the divergence of geodesics) even in spatiotemporal areas where no matter is present to be causally influenced.

Even if we grant that the metric features *cause* tidal effects, it does not follow that they are dispositional. Fundamental categorical features may also have a causal influence mediated, however, by laws of nature. The difference is that only fundamental dispositional properties are constituents of states of affairs which are *by themselves* (parts of) minimal truthmakers for modal truths which may describe causal powers. Yet, that is not the case with the putative causal impact of spacetime on matter. For, as Bartels (2013, 2008) himself clearly points out, metrical features

[30] For a presentation of the alternative interpretations, see Lehmkuhl (2008).

[31] This relative acceleration is directly related to the curvature of spacetime through the geodesic deviation equation. For details, see Wald (1984, 67).

produce their effects (i.e., 'causally' affect material bodies through measurable tidal forces) *only in conjunction with a lawful assumption*, the geodesic hypothesis, which ensures the compatibility of the affine structure of spacetime with the metrical structure. Acknowledging that, however, is tantamount to conceding that metrical features are *categorical*. The important issue for our purposes is *not* to distinguish (as Bartels does) *powers* from dispositional properties by ascribing metaphysical necessity to the results of the former (but not of the latter). The main issue, instead, is to differentiate between dispositional and categorical features. This distinction can be achieved via my preferred @-criteria of dispositionality/categoricality and has nothing to do with the metaphysical necessity of their results.[32,33]

1.3.1.5 A Nomic Interpretation of Bird's Counterfactual?

As I have already noted, there can be an alternative, *nomic* interpretation of Bird's counterfactual. Perhaps, the dispositional essence of spatiotemporal relations could be expressed by their essential nomic role in Einstein's law, in spite of the *non-causal* character of the latter. In other words, it might be argued, there is no need for the metric to have a causal essence in order to be construed dispositionally; a nomic counterfactual role is sufficient to provide the metric (and, implicitly, the spatiotemporal relations) with dispositional essence. In opposition to the above claim, I'll now challenge the hypothesis that dispositionalists can effortlessly think of Einstein's law as providing spatiotemporal relations with a dispositional essence. Recall that Einstein's equations express a global law of co-existence between space–time structure and mass–energy distribu-

[32] As far as I can see, there is nothing (except, perhaps, lack of strong motivation to embrace the relevant view) to exclude the possibility of metaphysically *necessary* laws between *categorical* properties. (In fact, Evan Fales (1993, 140) tentatively suggests such a view.) In this sense, the metaphysical necessity of dispositional properties' results could not be their definitional characteristic.

[33] To be fair, Bartels (2013, 2010) is clear that metrical features are not dispositional in the 'orthodox' sense. In fact, he argues for a *neutral monistic* account according to which 'if we stick to the mathematical definition of the metric tensor, we ascribe (physical) metrical properties in a categorical way, while if we characterize metrical properties by their capacity to determine tidal forces, we characterize them by their causal role, and thus as dispositions'.

tion of any (physically) possible world. In Ellis' (2001, 205) terminology, Einstein's law is like the global conservation laws or general structural principles. According to those philosophers (like Ellis) who believe in natural kinds' essences and laws that 'flow' from them, it is natural to construe such laws as characterising not natural kinds within the world, but the world as a whole (Bigelow et al. (1992, 384)). For them, therefore, even if Einstein's equations do express the dispositional essence of something, the only entity that this 'something' could be is the whole world. Nevertheless, the assumption that the actual world has a dispositional essence does not by itself imply that spatiotemporal relations have a dispositional essence as well. For we do not live in a purely spatiotemporal world (i.e., a world empty of matter) and so, the dispositional essence of the world may be plausibly due to the fundamental properties of material bodies and not to spatiotemporal relations. Dispositional essentialists need, in this case, a further argument in order to discard the above possibility.

Alternatively, according to those philosophers (like Bird) who hold that only *properties* give rise to laws, it is natural to construe global laws of co-existence as characterising the dispositional essence of a *world-property* corresponding to the kind of world we live in. They may argue in favour of Dispositional Monism by maintaining that the dispositionality of the world-property secures the dispositionality of spatiotemporal relations. To see whether this claim is tenable, let us suppose that there is such a global world-property and Einstein's equations describe its dispositional essence. Notice, first, that a dispositional monist cannot maintain that the aforementioned world-property is the only *fundamental* property of the world. For, in the opposite case, the whole discussion about the dispositionality of spatiotemporal relations (*qua* non-fundamental relations) is irrelevant to the viability of Dispositional Monism (and, *a fortiori*, so is the argument I present here). So, dispositional monists must acknowledge that, in addition to the fundamental dispositional world-property, there exist other properties and relations (including spatiotemporal ones) that are also fundamental. Given the fundamentality of spatiotemporal relations, let us now examine whether the alleged dispositionality of the world-property ensures that spatiotemporal relations are dispositional. As far as I can see, three are the most believable accounts of the ontological status of the global world-property. According to the first, the aforementioned global feature is a *structural* property that the whole actual world instantiates. By definition, then, the proper parts of the world instantiate natural properties and stand in natural spatiotemporal relations, which are all (properties and relations alike) constituents of the structural world-property. And there is a necessary connection between the instantiation of the structural property by the whole world and the instantiation of the constituent properties and relations by the parts.[34] In that interpretation, the dispositionality of the world-property does not assure that spatiotemporal relations are also dispositional. For we can follow a way of thought akin to the one we have previously followed (in the case of kind-essentialists) and plausibly maintain that the dispositional essence of the world-property may 'flow' not from the alleged dispositional essence of spatiotemporal relations, but rather from the dispositional essences of the other constituent fundamental properties (i.e., those of material bodies). Consequently, once again, dispositional essentialists need an extra argument to defend their claim.

The second option is to regard the world-property as an *emergent* property, distinct from any structural properties of the world. In that interpretation, the world-property is a novel, fundamental type of property possessing novel primitive causal powers. It is also a non-structural property, for its occurrence is not in any sense constituted by the occurrence of properties and relations of parts of the world. There are several accounts of the relationship between properties and relations of parts of the world and the emergent world-property. Emergence can be interpreted either as synchronic strong supervenience or as non-synchronic dynamical causal relation, or as a metaphysical relation called fusion (see, for details, O' Connor & Wong (2015)). Irrespective of interpretation, I see no reason to suppose that the putative dispositional character of the emergent world-property guarantees the dispositionality of the spatiotemporal rela-

[34] Though I talk about the *instantiation* of properties I do not presuppose that they are universals. Perhaps, both the world-property and its constituents are particularised features. In any case, nothing important for my purposes hinges on that. So the trope theorist may replace the verb 'instantiates' with the more neutral term 'having'. Furthermore, I do not take sides on whether the structural world-property is a first- or a second-order property of the actual world. For details about the ontology of structural universals, see Armstrong (1978), Lewis (1986) and Bigelow and Pargetter (1989).

tions from which it emerges. In any case, the onus is on dispositionalists to provide an argument for the above claim.

According to the final option, the world-property is ontologically akin to the *holistic* properties characterising quantum composite physical systems in entangled states (for details about the quantum case, see Karakostas (2004)). Besides, given that according to quantum mechanics the phenomenon of quantum entanglement is ubiquitous and most possibly (if quantum theory is true) establishes non-locality as a *fundamental* feature of the actual world, the existence of such a highly non-local, holistic fundamental natural property does not seem to be an unwarranted possibility. Now, to the extent that the world-property resembles ontologically the above-mentioned holistic features, it characterises the whole world but is neither supervenient on nor reducible to or derived from the properties of its parts and the relations they stand in. It is true that in the quantum case there is a sense in which a holistic property of a compound system determines the properties of its parts and their relations. For instance, consider a pair of spin 1/2 particles and their total spin. It can be shown that it is only the property of total spin (an intrinsic property of the compound physical system) that 'contains' all the information about the spin properties of the two particles. For only the entangled state of the whole system 'contains' the correlations among the spin probability distributions related to the particles (Karakostas (2004, 289–290)). However, despite the above remarks, there is no suggestion that the holistic property determines whether the properties of the parts are dispositional or not. So, the alleged ontological affinity of the world-property with the quantum holistic features provides us with no cogent reasons to assume that spatiotemporal relations 'inherit' the ontological feature of dispositionality from the global world-property.

To recap: Bird proposes that in order to prove that a fundamental feature has a dispositional essence we must find one specific counterfactual related to a causal or nomic role of the feature (as best science describes those roles) and then prove that a state of affairs involving that feature must by itself entail the aforementioned counterfactual. Spatiotemporal relations, however, fail to pass Bird's test. The suggested counterfactual cannot be interpreted as expressing a *causal* role of spatiotemporal relations.[35] If we alternatively suggest that it expresses an essential *nomic* role, the latter cannot be related to an alleged dispositional essence of spatiotemporal relations; it must rather be related to the essence of the entire world or to the dispositional essence of a global world-property. In both cases, the application of Bird's test does not ensure that the spatiotemporal relations themselves possess any kind of dispositional essence. Now, if Bird's suggested state of affairs involving the spatiotemporal relations cannot by itself be related to a 'proper' counterfactual in the actual world, it surely can neither by itself be a truthmaker for a 'proper' specific actual truth in order for the spatiotemporal relations to be dispositional in our world. Hence, the arguments of this section show that, according to my favoured criterion, the spatiotemporal metric relations cannot be construed as dispositional in the actual world. This, in turn, means that Dispositional Monism is a false view about the nature of *actual* fundamental properties and relations.

[35] This fact has also a bearing on the arguments of Bartels and Esfeld which are also based on the *causal* influence of spatiotemporal relations on the material world.

2

Against Identity Theory and Neutral Monism

2.1 Identity Theory

Identity Theory is based on the following core principle:

> *If X_i is a fundamental natural property of a concrete object, X_i is* ***simultaneously*** *dispositional and categorical; X_i's dispositionality and categoricality are not aspects or properties of X_i; X_i's dispositionality* ***is*** *X_i's categoricality and each of these is* ***identical*** *to X_i.*

It is a theory with a growing popularity among property theorists (Martin (1997); Martin and Heil (1999); Heil (2003); Strawson (2008); Schroer (2010); Jacobs (2011); Taylor (2013)). My conviction, however, is that the attractiveness of this view is mainly due to its promise to provide a solution to a (allegedly, in my view) serious problem of Categorical and Dispositional Monism that is, the (alleged) impossibility of purely categorical and purely dispositional properties.[1]

[1] An alternative solution to this problem is provided by the now largely neglected *Double-Aspect Theory*. According to this approach, each fundamental natural property has two *distinct*, though non-separable, aspects: the first contributes to the categoricality of its bearer, while the second to its

V. Livanios, *Science in Metaphysics*,
DOI 10.1007/978-3-319-41291-7_2

2.1.1 Problems for Identity Theory

2.1.1.1 The Objection from the Independent Variability of Dispositionality with Respect to Categoricality and Vice Versa

Let me begin by briefly examining an objection that, in my view, does not pose any serious difficulties for Identity Theory, at least in the context of fundamental properties. To illustrate the problem, consider any natural property characterising a concrete object. Identity Theory implies that you cannot vary its contribution to the categoricalities of the object without varying its contribution to the dispositionalities of the object and vice versa. It seems, however, that both directions of this claim can be challenged.

With respect to the first direction, the claim can be challenged by a *functionalist* account of the relation between categoricalities and dispositionalities. Functionalism implies that an object can be changed in respect of its categoricalities without being changed in respect of its dispositionalities. This is so because, from the functionalist point of view, dispositionalities are multiply realisable, higher-level properties possessed by an object in virtue of its possession of some distinct lower-level categoricalities. The challenge to Identity Theory emerges because we can have the *same* dispositionality with *different* categorical realisers. The identity theorist, however, can easily meet the functionalist challenge by claiming (as Heil (2004, 246–47) does) that what we actually have in cases of putative multiple realisability is a range of imperfectly similar properties (described by *one* vague predicate) rather than a single higher-level, multiply realised property. If that is true, no problem arises for Identity Theory. For each of the imperfectly similar properties, the identity claim of the theory holds. Each categorical 'realiser' is identical with a different dispositionality.

---

dispositionality. One may see the search for an adequate explanation of the postulated by Double-Aspect Theory *necessary* connection between the two aspects of any natural property as paving the way for the Identity Theory. For the latter relies inter alia on the conviction that the *best* explanation of the necessary compresence of the dispositional and the categorical is that they are *identical*.

It is not clear *how and to what degree* some properties might differ so that we may be entitled to both assume that they are different *and* describe them with the same non-specific predicate. But this is not important for our purposes anyway. Regardless of the adequacy of Heil's response, the functionalist objection does not threaten Identity Theory at the level of *fundamental* natural properties. Functionalism is a prima facie tenable hypothesis for non-fundamental dispositionalities such as those related to fragility or solubility. But it seems unmotivated for properties such as mass and charge. As far as I can see, we do not have any cogent reason to suppose that these properties are possessed by elementary particles in virtue of the possession of some distinct lower-level categoricalities. Hence, given that the ontological accounts under consideration in this book are primarily concerned with fundamental properties (a belief that many metaphysicians share), the functionalist challenge is not a serious objection to Identity Theory.

With respect to the second direction of the implication, the above claim might be challenged by invoking the possibility of worlds with alien laws. If there is such a possibility, then an object can be changed in respect of its dispositionalities without being changed in respect of its categoricalities. The *same* categoricalities in *different nomic* contexts might be related to different dispositionalities. The identity theorist might respond to this challenge by rejecting the possibility of worlds which, though inhabited by the actual properties, have alien laws. In particular, she may claim that the metaphysical necessity of actual laws (which is tantamount to the above rejection) is a natural consequence of the dispositional essence of fundamental properties. I am not sympathetic to this response, because I have reasons (which I'll present in Chap. 8) to deny the metaphysical necessity of laws of nature.[2] Nevertheless, the identity theorist might respond to the alien-laws challenge even if the possibility of alien laws is assumed. An example of such a response is given by Galen Strawson (2008, 277) who claims that the *totality* of actually possessed dispositions of an object includes the disposition to D_1 in nomic

[2] There are dissenting voices claiming that laws can be metaphysically contingent even if we grant that all fundamental properties have dispositional essences (see, for example, Hendry and Rowbottom (2009)).

context L_1, to D_2 in nomic context L_2, and so on. This totality does not change with the variation of the nomic context in different possible worlds. Strawson's suggestion may be interpreted as essentially ascribing to properties either a transworld *functional* essence or the same *total* essence, *part of* which is instantiated in each possible world. Under both interpretations, Strawson's tentative proposal certainly deviates from the orthodox account, according to which dispositional properties instantiate their whole, non-functional, essence in each world they exist. For me, however, this is no reason to dismiss it out of hand, especially since the orthodox account has its own difficult problems to handle. In conclusion, the arguments from the independent variability of dispositionality with respect to categoricality and vice versa do not pose any serious threat to Identity Theory.

2.1.1.2 The One Categoricality–Multiple Dispositionalities Problem

Consider the case of a spherical ball (Heil 2007, 16). In virtue of its sphericity, the ball would roll, would reflect light so as to look spherical, would make a concave impression in soft clay, and so on. It seems then that, according to Identity Theory, one property (sphericity), identical with a unique categoricality, is related to various distinct powers. If we *further* assume that each of these powers is related to a *different* dispositionality, then Identity Theory seems to imply the absurd fact that one and the same entity should be identified with several different entities. Sphericity is obviously a non-fundamental property, but it seems possible to extend the above conclusion so as to include the case of fundamental properties as well. For instance, in virtue of possessing electric charge, an electron could repel another electron, could attract a proton, could follow a specific path when entering a magnetic field, and so on. If we assume that electric charge is related to a *unique* categoricality, the difficulty alluded to above arises again. Is there an intractable problem here for the identity theorist?

I think not, because she may reject the assumption on which the above objection relies; namely, that each of the distinct powers of the object is related to a *different* dispositionality. She may, instead, hold the view that the dispositionality of any fundamental property is *multi-track*; that is, it is not associated with a *single* type of manifestation. The multi-track character of fundamental dispositionalities may be due to the qualitative or quantitative diversity of the stimulus conditions (for a recent discussion, see Vetter (2013)). Consider, for instance, the type of manifestation of the dispositionality of the negative charge of an electron. It depends on whether the stimulus is the negative charge of another electron, or the positive charge of a proton, or the presence of a magnetic field. Even if the stimulus is another negative charge, it depends on the magnitude of the stimulus-charge. A well-known metaphysical account supporting a multi-track interpretation is Martin-Heil's theory of reciprocal dispositional partners. According to this approach, what we have in the case of charge (and in the other cases as well) is a natural property related to a *single* dispositionality that *manifests* itself differently with different reciprocal partners. Some philosophers reject the multi-track interpretation (see, for instance, Bird (2007, 21–24)) insisting that each *fundamental* property should have a *single* manifestation.[3] The issue, however, is far from settled and, in the absence of conclusive arguments against it, the multi-track dispositionalities account is a live option for the identity theorist.

2.1.1.3 How Can We Understand (and Justify) the 'Surprising' Triple Identity?

The objections discussed in the previous two sub-sections are far from conclusive against the Identity Theory. In my view, however, the Identity Theory is an implausible metaphysical view of the nature of fundamental

[3] For instance, it might be objected that the multi-track account is supported by a notion of manifestation that erroneously makes the presence of any kind of stimulus (be it the instantiation of dispositional or categorical features) *necessary* for the manifestation of each property. For instance, Esfeld and Sachse (2011, Ch.2) suggest that the manifestation of the dispositionality of electric charge is not the attraction of another opposite charge but rather the spontaneous generation of an electromagnetic field. Thinking, falsely, that the manifestation of charge is the former type of event leads to the false belief that the presence of another charge in the vicinity of the first one is a necessary condition of its manifestation.

natural properties. The main reason is that it is based on a metaphysically obscure triple identity between a fundamental property and its dispositionality and categoricality. In what follows, I'll discuss three proposals to understand this identity and find them wanting.

Galen Strawson's Mediaeval Distinctions

Strawson (2008) starts off with reminding us the mediaeval categorisation of distinctions into *real* and *conceptual* ones. For him, 'a real distinction is a matter of what things *can* exist separately' (p. 271). He also thinks that the ground of inseparability of two entities (as we must interpret his term 'things' in order to cover the case under consideration) is identity. Therefore, he concludes that 'the fact or state of affairs that A = B is identical to the fact that there is no real distinction (but only a conceptual one) between A and B' (*ibid.*, 272). Given all that, Strawson's strategy for explaining the surprising identity of the categorical and the dispositional is the following:

(1) Dispositional and categorical 'aspects' of properties are inseparable.
(2) So, there is no *real* distinction between them; there is only a *conceptual* distinction.
(3) The fact, however, that there is no real distinction between the two aspects is *identical* to the fact that dispositional ≡ categorical.

Let us grant Strawson premise (1); it is a reformulation of the (alleged) fact that there is no possible world in which the dispositionality and the categoricality of any natural property exist separately. The main difficulty in Strawson's argument arises when he takes the step from (1) to (2): The fact that two entities can exist separately is only a *sufficient* but not a necessary condition for there to be a real distinction between them. As Oderberg (2009) convincingly claims, the necessary and sufficient condition for two entities to be really distinct is that *they* are numerically distinct, not only the concepts that we use for them or the terms with which we refer to them (2009, 678). Hence, the step from the inseparability of the dispositional and the categorical to their identity is illegitimate.

(Incidentally, if, *pace* Strawson, there is a real distinction between dispositional and categorical, what can explain their [alleged] inseparability? Oderberg's response is that, in all real distinctions between non-separable entities, it is the *essences* of the latter that best explain their inseparability (*ibid.*, 678). I do not intend to consider this view; the relevant for our discussion aspect of Oderberg's arguments is that Strawson has not provided us with a convincing explanation of the surprising identity.)

Martin-Heil's Duck/Rabbit Analogy

Martin and Heil invite us to consider an ambiguous figure such as the well-known duck/rabbit figure or the Necker cube. They think that the way we understand how we can consider in different ways these figures may shed light, *qua analogous*, on the case of the identity between dispositionality and categoricality. The considered shape is supposed to be analogous to the dispositionality or the categoricality of a property. Here is Heil:

> A property's dispositionality and its qualitativity are, as Locke might have put it, the self-same property *differently considered.* (2004, 243–44, my emphasis)
>
> [According to a misunderstanding of the identity view a] capacity for *partial* consideration presupposes distinctions in the world, and distinctions are a matter of differences in properties. If a property could be considered either as a disposition or as a quality then it must incorporate distinct features answering to these two modes of consideration. (*ibid.*, 250, my emphasis)

There are two distinct but interrelated issues which arise from the above quotations. First, there is the vital question concerning whether a way or a mode of consideration of a natural property presupposes distinctions in the world (in particular, ontological complexity of the property in question). And then, there is the *further* question whether the ontological distinctions (if needed) should be cashed out in terms of properties (in our case, properties of properties) or not. Heil explicitly rejects the view that our capacity for partial consideration of properties (which grounds the surprising identity claim) presupposes an ontological distinction. This belief, of course, entails that he also holds that there are no distinct *features* (namely, properties) of properties answering to the different modes of their consideration. Heil offers an independent defence for the latter view by providing an explanation for the alleged plausibility of the belief expressed in the above second quotation. He thinks that this belief is only justified when one misunderstands the context of discussion and considers the case of properties as analogous to the case of concrete objects. For him, contrary to the latter case where considering the same object in different ways is often a matter of considering different *properties* possessed by the object, in the case of properties we do not have to appeal to properties because what we consider is the *whole* property in different ways. Of course, Heil's insistence on the lack of ontological complexity corresponding to the different ways of considering properties should not come as a surprise. If the dispositionality and the categoricality of any natural property were somehow related to either features or 'parts' of it, then the core identity claim of Martin-Heil's theory between the property itself and its dispositionality and categoricality would obviously be compromised.

Where does this leave us concerning the interpretation of Identity Theory? Following Taylor (2013) and Schroer (2013), I think that Martin-Heil's analogy can be construed as an application in the case of properties of a kind of the *partial consideration strategy* (see, for a justification of this claim, the above second quotation). Namely, we can understand Martin and Heil as claiming that what were formerly thought of as distinct entities are actually *one* entity, partially considered. By definition, to partially consider a natural property (as either dispositional or categorical) is to attend to the property under one concept (let us say, the dispositional one), while ignoring the possibility of considering it under the alternative, *complementary* concept. Is this interpretation a plausible one? As Taylor (*ibid.*) remarks while explaining Schroer's approach, so long as both concepts of categoricality and dispositionality are 'thick' enough so that they can appear as corresponding to genuine and separable entities in the world, Identity Theory is surely in trouble. Schroer (*ibid.*) tries to meet this difficulty by presenting an extremely 'thin' conception of categoricality as a primitive 'something' common to all concrete objects occupying space that *just* makes them different from empty space. This 'poor' nature of categoricality (which, incidentally, is analogous to the 'thin' conception of the substratum in the case of concrete objects) might be seen as vindicating the claim that the concepts of the dispositionality and categoricality do not after all correspond to separable entities. I discern two problematic aspects in Schroer's strategy. The first (the less problematic) has to do with the incentive behind the specific proposal about the 'thin' nature of categoricality. Schroer justifies his suggestion on the grounds that it can provide an adequate solution to what he thinks is the major objection to Dispositional Monism; that is, the accusation that, if all properties are pure dispositional, there can be no difference between 'concrete' objects and empty space. Schroer argues that, by embracing the view that each natural property has a categoricality which *merely* bestows the capability of space occupancy, he can meet this problem. His choice, however, is justified provided that another objection (the one which, in my view, almost monopolises the current discussion about the plausibility of Dispositional Monism), that is, the regress-of-powers objection,[4] can be successfully met either by Schroer's proposal or by independent means. But as far as I know, the jury is still out on that issue (see, for instance, Swinburne (1980); Blackburn (1990); Holton (1999); Lowe (2006); Bird (2007a); Bigaj (2010); Oderberg (2011; 2012); Shackel (2011); Ingthorsson (2012)). The second aspect is more problematic and concerns the very efficiency of the suggested account. If the intended target is to show that the identity of dispositionality and categoricality is less incredible than previously thought, it is far from clear that it can be achieved by arguing for the inseparability of these two 'aspects' (recall Oderberg's arguments against Strawson's view). Hence, if inseparability is what Schroer attempts to argue for, his attempt does not enhance the plausibility of Identity Theory.

Be that as it may, the basic difficulty with the partial consideration approach (where what we actually have is an act of mental abstraction) is

[4] According to Bird's (2007) influential account, the most serious version of the regress-of-powers-objection concerns the identity of dispositional properties in a dispositional monistic world. Roughly, if the identity of any dispositional property is determined by its relations to other properties, and those other properties are dispositional as well, then either there is an infinity of properties or there is circularity in this relationship of identities.

that we come dangerously close to the neutral monistic view according to which properties *in themselves* are neither dispositional nor categorical. This result, besides nearly collapsing Identity Theory to Neutral Monism, threatens the very coherence of the identity account; under the neutral monistic perspective, properties cannot be identical to their dispositionality and categoricality since, by themselves, are neither dispositional nor categorical.

Jacobs' a Posteriori Identity

Jonathan Jacobs (2011) presents an alternative understanding of the core principle of Identity Theory. To this end, first, he proposes that to be categorical is to be identical with a *thick quiddity*, where the latter is conceived not as an ontological *constituent* of a property, but as the property itself.[5] Then, he suggests that to be dispositional is to be a nature sufficient to be (a part of) the truthmaker for certain counterfactuals. Finally, he claims that there is (a non-surprising!) identity of the categorical and the dispositional which is analogous to Kripkean *a posteriori identities* such as between Hesperus and Phosphorus. As Jacobs himself admits, for the analogy to work, we have to find a way to distinguish dispositional from categorical in such a way that, though we can imagine that they might be distinct, they are in fact identical. Given the achievement of this task, the identity theorist could either deny that the conceivability of distinctness implies metaphysical possibility or accept that it does but misdescribes it.

To evaluate Jacobs' proposal, two points need to be stressed to start with: first, Jacobs refers to the conceivability of *distinctness* between dispositionality and categoricality, not of their *separability*. Hence, his approach is prima facie less problematic than Schroer's which refers to the non-conceivability of their separate existence (which, most probably, does not imply their identity). Second, he does not want to show that categoricality, due to the 'thin' conception of it, cannot be conceived as separable from dispositionality. He identifies categoricality with thick

[5] According to one interpretation, a quiddity of a property is nothing other than the property itself, whereas according to an alternative construal, it is an intrinsic aspect of the property, that is, a second-order property. For more details, see Sect. 4.2.

quiddity, leaving relatively unexplained the proper conception of the latter. This move, of course, leaves room for a 'thick' enough conception of categoricality; this conception, however, can correspond to an entity which (*pace* Schroer) can be justifiably a posteriori identified with another entity (dispositionality) characterised by a completely different (and 'thick' as well) conception.

Despite the above comparative advantages, however, Jacobs' account faces a crucial difficulty. To highlight that, we must first note that it is ultimately a *brute* fact, 'forced' by the a posteriori identity of the dispositional with the categorical, that a thick quiddity is also sufficient to make true certain counterfactuals.[6] What Jacobs attempts to do is (as he says) to distinguish the dispositional from the categorical in such a way as to make us capable of imagining them as distinct, though they are actually identical. But unfortunately, that's not enough; what are the a posteriori grounds for the identification of the dispositional with the categorical? In other words, what is it that makes us think that the *same* entity is the truthmaker for these *different* truths?[7] The only reason I can envisage is the a posteriori discovery that each natural property has simultaneously a categorical and a dispositional 'character'. Examples (though controversial) abide in the realm of non-fundamental properties of macroscopic objects. Consider, for instance, the example of a spherical ball capable of rolling down an inclined plane. It seems that geometrical properties of macroscopic objects, such as sphericity, are both 'qualitative' (categorical) and confer (by themselves?) causal powers on their bearers. The case of fundamental properties, such as mass, charge and spin, is more subtle. It is true that those properties seem to have 'dispositional' characters related to the fundamental interactions associated with them. It is also true that one may find candidates for 'categorical' characters of fundamental properties in other aspects of them (such as their *mathematical* characterisations). However, even if we grant all that, why don't we assume that each property has two dis-

[6] The theory of truthmakers allows doing that without being ontologically structured. A truthmaker is not required to be structured in the way the truth it makes true is.

[7] Note once again that the problem is *not* that we have one truthmaker for more than one truths; the truthmaking relation is not necessarily injective. It rather concerns the grounds we may have to assume that this is the case here.

tinct but non-separable truthmaking 'aspects' which make true the categorical and the dispositional truths concerning properties accordingly? What is the extra reason to identify these 'aspects' with the property itself? As far as I can see, Jacobs does not provide one.

2.1.2 A Unique Categoricality?

I conclude the examination of the prospects of Identity Theory by briefly discussing Robert Schroer's novel form of the theory which inter alia promises to meet the difficulties besetting it. His strategy is to challenge the tacit assumption of most property theorists according to which, if there are *any* categorical properties in the world, then there is a *multiplicity* of them. Schroer (2010) rejects the view that there are many *determinate* ways of being categorical and argues for the possibility of existence of *only one* categorical property (as I have already remarked, this unique categorical property[8]contributes 'something' to the objects instantiating it and renders them different from empty space). There are two interpretations of Schroer's proposal which avoid the obvious triviality of considering his unique categoricality as the highest *determinable* feature-of-categoricality which by definition all properties possess. According to the first, *each* absolutely determinate property of *each* determinable quantity should have the same categorical aspect. The second construal leaves room for a diversity of determinate categorical aspects of a *single* determinable categorical aspect which *all* kinds of properties 'share'.

Schroer's account has several advantages compared to the other, more 'orthodox', versions of Identity Theory. First, it may in a sense relieve the tension created by the surprising identity when it 'explains' why specific dispositionalities are always accompanied by the same categoricalities. The necessary identity between the various dispositionalities and the unique categoricality seems to be more palatable because, inter alia, it is posited to account for the obvious difference of concrete objects from empty space and not for the variety of the actual natural properties.

[8] Schroer, *qua* identity theorist (see Schroer (2013)), does not believe in the existence of distinct kinds of property. For him, any natural property is both categorical and dispositional.

Second, the account provides an obvious solution to the functionalist challenge. Dispositionalities cannot be multiply realisable simply because there is no variety of categoricalities to realise their functional role. Finally, the alien-laws challenge does not prima facie raise any difficulty for Schroer's account. Not only is the possibility of the same unique categoricality related to different dispositionalities (in different nomic contexts) ensured, but also this becomes the standard situation even in the actual world. This result seems prima facie innocuous, to the extent that there is a necessary connection between a given property and the powers the property bestows on its bearers (but it leads, as we will soon see, to contradiction if we take into account the core claim of Identity Theory).

Though prima facie promising, Schroer's account raises a difficulty (presented by Taylor (2013) and generalised here) which I think, regardless of the adequacy of the responses to the other possible worries, is fatal. Roughly, the objection is this: According to Martin-Heil's original version of Identity Theory, there is a variety of categorical and dispositional 'natures' which, through the core identity claim of the theory, yields a variety of natural properties. Schroer also wants to defend a version of Identity Theory which, unlike the original one, admits a unique categorical 'nature'. But this assumption, combined with the 'surprising' identity of Identity Theory, has unpalatable consequences. Suppose first that we follow the first interpretation of Schroer's unique categorical 'nature'. Then his view has the absurd consequence that *all* determinate natural properties, regardless of the determinable which they fall under, are identical to this unique 'nature' and, so, are themselves identical as well. A slightly weaker (but equally disturbing) conclusion follows if we alternatively follow the second interpretation. In that case, there is only *one* kind of determinable physical quantity in the world; it is the one which is identical to this unique 'nature'.

To sum up, in this section I argued that identity theorists have not provided us an adequate theoretical explanation of the suggested triple identity between a fundamental property, its dispositionality and its categoricality. This is the case in both the 'orthodox' many-categoricalities versions and Schroer's unique-categoricality form of Identity Theory (the latter, as I have argued, leads to absurd consequences as well). In the absence of such an explanation, I conclude that Identity Theory, in spite of its initial attractiveness, is not a viable metaphysical account of the nature of fundamental properties of our world.

2.2 Neutral Monism

According to Neutral Monism, the dispositional/categorical distinction carries *no* ontological weight. It is a distinction concerning the *predicates* we use to describe fundamental properties rather than their distinct ontological kinds. Adherents of this view are early Mumford (1998) and Mellor (2000). In Mumford's version, all fundamental natural properties can be characterised either dispositionally or categorically, relative to a particular causal role. If we want to describe a property and choose a predicate which has a conceptually necessary link with the particular causal role, then this predicate represents this property as dispositional. But, if we choose a different predicate for the property which *doesn't* conceptually necessitate the particular causal role, then we represent the same property in categorical terms.

The neutral monistic view about properties is clearly motivated by the fact that *all* fundamental natural properties can be characterised by their relationship to certain counterfactuals. Given that, neutral monists claim that we have no cogent reason to assume that there are multiple kinds of properties in the world. Nevertheless, there is an intuitive distinction (at least as far as non-fundamental natural properties are concerned) between dispositional and categorical features which must be accommodated. Hence, neutral monists must insist (as they in fact do) that the distinction is merely *conceptual*; we have just one kind of natural property described, nonetheless, in two different ways. But as far as I can see, this motivation is rather weak since we can reach an alternative conclusion via the above reasoning. In particular, we may plausibly claim that the aforementioned relationship to counterfactuals is not an adequate criterion for the categorical/dispositional distinction, which, nevertheless, marks an *ontological* distinction between two kinds of properties.

Neutral Monism is not a popular view nowadays. In Sect. 2.2.2, I'll present my own reason to reject it. Meanwhile, in what follows, I'll examine an argument that prima facie shows that Neutral Monism is untenable.

2.2.1 Neutral Monism and the Modified Ungrounded Argument

Mumford (2006) presents an argument—the so-called *Ungrounded Argument* (UA[9] for short)—against the view that every *dispositional* property is grounded in some property other than itself.[10] The argument runs as follows:

(1) There are subatomic particles which are simple.
(2) That which is simple has no lower-level components or properties.
(3) The properties of subatomic particles are (all) dispositional.
(4) The grounds of a dispositional property can be found only among the lower-level components or properties of that of which it is a property.
∴ The dispositional properties of subatomic particles have no ground and, by existential generalisation, there exist some ungrounded dispositions.

UA *presupposes* (premise 3) that all state-independent properties of elementary particles (such as rest mass, charge and spin) are dispositional. Mumford (in line with Ellis (2001)[11] and Thompson (1988)) claims that this assumption is supported by physical theory (contemporary physics)

> not just as it is interpreted by philosophers, but also by scientists disinterested in this debate. (2006, 475)

Though I have strong reservations about the alleged support by science of the *ontological* claim that all fundamental properties are dispositional, here I just point out that, by *assuming* the existence of dispositional prop-

[9] For a detailed critique of UA, see Williams (2009).

[10] The conclusion of the argument is compatible with McKitrick's (2003) view. She argues for the possibility of *bare* dispositions that have no *distinct* causal basis. For McKitrick, the causal basis of a bare disposition is the disposition itself (in the sense that it causally explains its own manifestation).

[11] Mumford is a realist about powers, but, to my knowledge, he is not convinced about an Ellis-like essentialism about natural kinds.

erties, UA cannot be regarded as an argument against Neutral Monism. Consider, however, the following argument (let us call it MUA) which *can* be interpreted as offering sufficient reasons against the neutral monistic view.

(1) There are subatomic particles which are simple.
(2) That which is simple has no lower-level components or properties.
(3) The properties of subatomic particles *can be characterised* dispositionally.
(4) The grounds of a natural property can be found only among the lower-level components or properties of that of which it is a property.
(5) Given (1) and (2), the properties of subatomic particles are ungrounded.
(6) A categorical description of properties of subatomic particles can be given only if they are grounded.
(7) Given (5) and (6), the properties of subatomic particles can be characterised dispositionally but not categorically.
∴ Neutral Monism is false.

Let us first focus on premise (3). The dispositional description of state-independent properties seems to be explicitly related to their (theoretically described) power to produce *by themselves* specific phenomena. This is due to the fact that, given the truth of premise (5), the (causal) activity of these properties cannot be based on a (categorical) basis. Though it is debatable whether such a dispositional description uses predicates which *conceptually necessitate* the actual (causal) roles, I proceed as if it actually did. MUA is clearly valid, but in my view, it is not sound since premise (6) is false. It is simply not true that properties of subatomic particles can be given a categorical description only if they are grounded. Contemporary science informs us that, besides the 'dispositional' description through causal roles, there is an alternative description of the state-independent ungrounded properties Mumford invokes which is independent (a) of their causal roles, and (b) of any reference to entities (or kinds of entity) that may instantiate those properties.[12] As a matter of fact, *state-independent fundamental properties can be characterised as invariants under the action of certain fundamental symmetry groups.* Here are some necessary details. The notion of symmetry is closely related to the concept of invariance (of an object or a law of nature). Symmetry means that there are transformations which leave specific features of objects or laws invariant. A particular symmetry can be identified, provided that knowledge of (a) the specific transformation and (b) the specific feature left unchanged has been acquired. There are two important distinctions characterising the symmetries of interest in physics: internal/external and global/local symmetries. An external symmetry is one in which the corresponding transformation amounts to a variation of a spatiotemporal variable. For instance, spatial rotations, spatial and temporal translations and boosts are all transformations associated with external symmetries. By contrast, an internal symmetry is one in which the corresponding transformation does not mean a change of spatiotemporal variables.[13] The second distinction concerns the dependence of symmetries on space–time parameters. Symmetries are global if they are not functions of space–time variables. They are local, if they are.

It is standard mathematical procedure to discuss symmetries in the context of group theory. Various transformations of physical interest form groups which can be analysed mathematically. Given that the fundamental equations of a theory describing a physical system are invariant under the transformations of a specific group, the states of the system transform into each other according to some representation of the group. Wigner (1939) showed that the states of elementary particles transform according to irreducible representations[14] of a space–time group called

[12] My intention is to exclude certain trivial 'categorical' descriptions. For instance (arguably), one possible 'categorical' description (in the sense of independence of causal roles) of the property 'electric charge' is that it is a property possessed by electrons, protons and other charged particles.

[13] Variation of the phase of the wave function of a particle is an example of a transformation associated with an internal symmetry.

[14] By definition, a *representation* of a group on a vector space is an action of the group on the space, in which each member of the group acts as a linear transformation. A subspace of the vector space is said to be *invariant* if, for each vector of the subspace and each member of the group, the action of the group element on the vector yields another vector of the subspace. A representation is called *irreducible* if the only invariant subspaces of the vector space are the space itself and the subspace containing only the null vector.

the Poincaré group.[15] He computed all irreducible representations of the Poincaré group on the space of states of elementary particles, and found that all representations satisfying certain physically reasonable conditions can be labeled by two parameters m and s, where m can be identified with the rest mass of particles and s can be identified with their spin.[16] Hence, two fundamental properties, rest mass and spin, which according to dispositional essentialists[17] are dispositional features of elementary particles, can be identified via a symmetry-based procedure independent of their causal/nomic roles.

Another fundamental physical property which is also popular among dispositional essentialists and can be identified via symmetry considerations is electric charge. It has been shown that there exist transformations of a specific internal continuous symmetry (the corresponding group is called U(1)) which leave invariant the Lagrangian function of any (com-

---

[15] The Poincaré group (aka inhomogeneous Lorentz group) is associated with the global external symmetry under the action of Lorentz boosts and rotations, and of space–time translations.

[16] According to group theory, associated with any irreducible representation of a continuous group are operators (called Casimir operators) which are multiples of the unit operator, and, therefore, commute with all operators in the representation. Casimir operators have fixed numerical values in a given irreducible representation, which can be used as labels characterising the irreducible representation (Hamermesh 1989, 318). The Casimir operators of the Poincaré group are

$$C_1 = -m^2 \quad C_2 = -m^2 s(s+1)$$

Rest mass is the property which appears (as a parameter) in the first Casimir operator; that is, the one which is formalised with the aid of only one parameter (which, of course, represents mass). Having identified mass, we can then identify spin as the property represented by the second parameter which appears in the second Casimir operator (the one which is formalised with the aid of two different parameters).

[17] Dispositional Essentialism is a species of Dispositional Realism which primarily concerns the nature of fundamental natural properties. According to Ellis and Lierse (1994, 39), it is a metaphysical position which 'is realistic about the dispositional properties of the fundamental particles and fields and essentialist for two reasons: first, because it holds that these properties are among the essential properties of these particles and fields and, second, because it holds that it is essential to the natural processes in which these particles and fields may be involved, that they should be displays of these dispositional properties'. Dispositional Essentialism does not hold that all fundamental natural properties are dispositional. According to Bird's brief definition, Dispositional Essentialism is the metaphysical view that holds that at least some sparse, fundamental natural properties (and relations) have dispositional essences (i.e., have some essence that may be characterised dispositionally).

plex scalar) field.[18] Furthermore, there is a well-known theorem (Noether's first theorem[19]) which connects continuous symmetries, conservation laws and conserved quantities. The application of Noether's theorem in the case of the aforementioned symmetry yields a conserved quantity which can be identified with electric charge (Ryder 1985, 93).

After the brief technical digression, the crucial point: I take it that the predicate 'being invariant under the action of a group associated with symmetry X' does not conceptually necessitate any particular causal role; therefore, by Mumford's own lights, it offers a categorical[20] description of properties such as charge, mass or spin. It seems then that the 'broken' symmetry between categorical and dispositional descriptions, which is at the core of Neutral Monism, might be restored.[21]

It might be objected that the suggested description via the epistemic symmetry-based identification concerns only *determinable* properties; it is not possible (following the suggested method) to identify *specific values* of any fundamental property. So it seems that, in order to describe and identify a specific *determinate* property falling under the determinable one, we should certainly use the various roles which that property is capable of playing in specific circumstances. Associating this remark with the acceptance of the *necessity* of epistemic individuation of deter-

---

[18] In technical terms, the Lagrangian function L = T − V (where T and V are the kinetic and potential energy respectively) of any massive scalar field φ is invariant under the action of the following transformations:

$$\phi \rightarrow e^{-i\Lambda}\phi \qquad \phi^* \rightarrow e^{i\Lambda}\phi^*$$

where Λ is a real constant and φ^* is the complex conjugate of the field φ. These transformations are associated with the group U(1) and the corresponding (global) symmetry is an internal one.

[19] Consider the Lagrangian density L of a physical system and the action S related to it. According to Noether's first theorem, if the action is invariant under a continuous group of transformations depending smoothly on independent constant parameters, then, given that the equations of motion of the system are satisfied, there are continuity equations for currents associated with each parameter on which the symmetry group depends. Given appropriate boundary conditions, each continuity equation corresponds to a conserved quantity. For technical details, see, for instance, Ryder (1985, 87–93).

[20] To avoid misunderstandings, by stating this conclusion, I do not endorse Mumford's conceptual criterion for the dispositional/categorical distinction. I just show that, by Mumford's own lights, his version of Neutral Monism has nothing to fear from the existence of ungrounded dispositions.

[21] For a response to the obvious objection that the suggested characterisation is *not* independent of causal roles after all, see the relevant discussion in Sect. 4.2.

minate properties would turn it into a possible objection to my suggestion. Seeing the issue of individuation in such a light is compatible with any thesis which argues against the existence of determinables as genuine properties. However, it is also compatible with any view accepting the existence of genuine determinable properties, while denying at the same time the possibility of their identification independently of the identification of determinates falling under them. In the following remarks, I presuppose the existence of genuine determinable properties, although the final conclusion seems to be independent of this condition. I shall try to show that, given the basic features of the determinable–determinate relation, my suggestion provides a means for the individuation of the determinables and the *partial* identification of determinates which fall under them.

Considering the determinable–determinate relation, we find that the most important truism about it is this: (An object) having a determinate property implies its having a determinable property in a *specific* way. In general, the features according to which a determinable can be determined are called (following Funkhouser's (2006) terminology) *determination dimensions* of the property. Determinates that fall under the same determinable can differ only with regard to their values *under the same determination dimension.* Yet there are other features (called, by Funkhouser, *non-determinable necessities*) that each determinable and determinate must have, but which do not admit of variation. Determinates that fall under the same determinable cannot differ with regard to their non-determinable necessities. For instance, various isosceles triangles (interpreted as determinates of the determinable 'being a triangle') can differ with regard to the length of their sides, but not in their three-sidedness (Funkhouser 2006, 551). Let us now consider the determinable property 'mass' and the determinate property 'mass 5 kg' that falls under it. I think that we can plausibly claim that all the determinates that fall under the determinable 'mass' differ only with regard to their values under the one determination dimension corresponding to the scalar quantity 'mass'. They do not, however, differ with regard to their invariance under the Poincaré group. Invariance under that global symmetry group is the non-determinable necessity in the case of mass. Using this feature (which all determinates of 'mass' share), we can *partially* identify any determinate that falls under the determinable 'mass'. Surely, this method does not provide a *complete* identification of each determinate property, since it does not determine the exact value under the determination dimension corresponding to the quantity 'mass'. Nevertheless, the application of the method clearly indicates the possibility of an independent (from the prior identification of determinates) procedure for the individuation of the determinable property, *through which a partial identification of determinates can be achieved.* I think that this conclusion holds even if we deny the existence of genuine determinable properties. In this case, a determinable 'property' is nothing but a set of genuine determinate properties which 'resemble' one another in some respect; a reasonable account (though not the only one) of what it is that unifies that set into a resemblance class[22] is my suggestion: the invariance under the action of fundamental symmetry groups.

Of course, the general question remains: Are there equally acceptable categorical and dispositional descriptions for *all* genuine natural properties? Any attempt to give an answer to this question is plagued by several difficulties. First, how are we to define genuine properties? Even if we admit the existence of only sparse properties, we do not know which properties can be classified as such. Physicists posit certain fundamental properties to explain physical phenomena, but do they exhaust the range of genuine natural properties? Perhaps other properties (posited by other sciences, or acknowledged by 'common sense') are genuine, too. Second, even if we decide that the only genuine properties are ultimately those which physicists posit, the preceding discussion is still incomplete (for instance, we have not examined the case of the *state-dependent* properties of elementary particles, such as momentum, angular momentum, etc.). Nevertheless, what the preceding remarks clearly show is that (at least as far as state-independent properties are concerned) premise (6) of MUA is false. Hence, MUA cannot be considered as a fatal blow to Neutral Monism (at least to Mumford's version of it).

[22] Of course, there is always the option to take such a resemblance as a brute fact.

2.2.2 Agnosticism

In my view, the most serious problem that besets any version of Neutral Monism is this: On the one hand, neutral monists claim that property realists do not have to commit themselves to the existence of *distinct kinds* of property. They insist that there actually are fundamental properties *simpliciter*, and this claim is clearly an ontological one. On the other hand, the only thing that they can say about the nature of these natural properties is that they are *such* that they are representable by two different epistemic descriptions. The obvious ontological question that arises is, what must be the nature of the actual fundamental properties in order to 'fit' our different descriptions? The question is pressing but neutral monists take an *agnostic* stance on this issue, claiming that any ontological commitment about the nature of properties is unwarranted. This agnosticism must accept the possibility of different ways of description as a *brute* fact about properties. Otherwise, in any attempt to give an ontological explanation of this possibility, the above question arises again. Now, the problem is *not* that neutral monists refuse to accept the existence of an ontological correlate of a conceptual distinction. This would perhaps be justified were the conceptual distinction to play a minor role in our ontological considerations. The dispositional/categorical distinction, however, is intimately related to the two major ontological worldviews: the Humean and the Non-Humean. Therefore, I find it very difficult to understand how a metaphysician who takes seriously the debate between the two worldviews (and I am sure that the overwhelming majority of property theorists do) may nevertheless refrain from investigating its ontological grounds in the nature of fundamental properties. Hence, I think, every property realist has a definitive reason for rejecting the agnostic stance of Neutral Monism.

Can, at least, the ontologist investigate which ontological frameworks are *compatible* with the neutral monistic view? This is far from easy, especially if she endorses Mumford's criterion of the conceptual dispositional/categorical distinction. I assume that all metaphysicians accept that the dispositional description of fundamental properties is *compatible* with non-Humean metaphysics (i.e., the view according to which there are real, active, occurrent dispositional properties in the world, and laws of nature are supervenient on them). But if we take seriously the basic tenet of Neutral Monism, we should also take into account the categorical descriptions of fundamental properties. Is the categorical description compatible with a Humean or a non-Humean picture of the world? It is not possible to give an answer to this question, unless we somehow *delineate* the appropriate kind of categorical description. The suggestion I presented above concerning Mumford's version of Neutral Monism offers a defensible option as far as fundamental state-independent properties are concerned. In that case, however, the answer to the compatibility question clearly depends on the *interpretation* of symmetries. Let me briefly show why. According to a first kind of position (*ontological* interpretation), symmetries represent properties existing in nature (or properties characterising the structure of the physical world). The physical existence of symmetries explains the success of symmetry considerations in modern physics (Castellani 2002, 191). From an alternative point of view (*epistemic* interpretation), however, the presence of symmetries in physical theories is related either to some general conditions on the nature (or the possibility) of physical knowledge or to some limits inherent in our way of describing the physical world (*ibid.*, 192). Finally, symmetry principles are often taken to be conditions for the possibility of a physical description in terms of laws, or interpreted as *transcendental* principles in the Kantian sense.

Now, if the epistemic version is adopted, the ontological picture is not constrained by any means. If, for instance, symmetries are *prerequisites* for the possibility of the cognitive access to fundamental properties, any ontological account is compatible with the categorical descriptions of those properties. The fact that we describe the properties as invariants under the action of fundamental symmetry groups does not provide us with any information about the nature of the properties themselves. But if, on the other hand, the ontic interpretation is adopted, an account of the symmetries *themselves* is needed, in order to investigate the compatibility of the categorical description with a suggested nature of a state-independent property. In this case, symmetries, being as real as all other properties of the world (although, perhaps, second-order ones), become part of the problem rather than a means to its solution. The debate about the interpretation of symmetries is far from being settled. Therefore, for the time being, it would be wise to refrain from answering any questions concerning the compatibility (of my suggested version) of Mumford's early neutral monistic view with the two candidates for the nature of fundamental (state-independent) properties of elementary particles. Of course, this conclusion does not mean that we should stop exploring whether actual fundamental properties are dispositional or categorical (or both). We just have to reject Mumford's conceptual criterion, follow an *ontological* criterion of the dispositional/categorical distinction and check whether (according to it) *all* fundamental properties satisfy the criterion of dispositionality (categoricality) or not. This is the strategy I've followed throughout the present work in order to defend (in the next chapter) the view that our world is a categorical–monistic one.

3

In Defence of Categorical Monism

The definitional tenet of Categorical Monism is the following:

All fundamental natural properties are purely categorical.

Categorical Monism has been challenged both directly (through argumentation for the genuine existence of actual fundamental dispositional properties) and indirectly (by stressing the 'unpalatable' consequences of adopting the existence of fundamental categorical properties). In Chap. 1, I implicitly responded to the latter challenge by arguing that there actually exist fundamental categorical relations. Yet, this response is at best sufficient only for a defence of a kind of *Property-Dualism*[1]; for, all that (I hope) it shows is that in any case we have to allow for the existence

[1] Property-dualists (henceforth, dualists) argue for the actual existence of two ontologically distinct and mutually exclusive *kinds* of fundamental property. More precisely, the core principle of Property-Dualism is this:

The members of a subset of the set of all fundamental properties are exclusively categorical, while the members of its complement (with respect to the set of all fundamental properties) are exclusively dispositional.

Philosophers who advocated dualistic theses are Prior, Pargetter and Jackson (1982), Prior (1985), Place (1996), and, more recently, Ellis (2001), Molnar (2003) and McKitrick (2009).

V. Livanios, *Science in Metaphysics*,
DOI 10.1007/978-3-319-41291-7_3

of some actual fundamental categorical features without, however, arguing that there are no actual fundamental dispositional features. Here I wish to argue in favour of the categorical monistic view. To this end, I'll present a novel argument which purports to show that actual fundamental properties *cannot* be dispositional.

3.1 In Defence of the Categoricality of Fundamental Properties: The Argument from Renormalisation

Contrary to claims of the opponents of Categorical Monism, I think that there is an argument from scientific practice strongly suggesting that all fundamental properties in the actual world are categorical. The argument is inspired by the *renormalisation methods* used in the context of Quantum Field Theories (QFTs, for short).[2] The QFT framework is the most fundamental and empirically confirmed theoretical framework we currently have for describing the fundamental interactions occurring in nature. Each interacting QFT describing the specific local interaction between fields is characterised by its own Lagrangian function which contains a number of parameters called *coupling constants*. According to the QFT framework fundamental forces occur via the exchange of bosons, and coupling constants enter into the calculations of the probability of emission and absorption of these exchange bosons during particle interactions. Furthermore, the strength of each fundamental force depends both on the mass of the exchange bosons that 'carry' the force *and* on the actual value of the relevant coupling constant.[3] The enormously accurate predictions achieved by QFTs are related to the scattering amplitudes associated with specific processes. Due to the complexity of the calculations involved, theoretical physicists employ perturbation theory

[2] The renormalisation techniques are described in every textbook on QFT. See, for example, Peskin and Schroeder (1995).

[3] In the unified framework of the Standard Model of fundamental interactions, there exist specific relations among coupling constants. For example, in the electroweak sector of the Standard Model, the weak $SU(2)_L$ coupling g, the weak U(1) coupling g′ and the electric charge e are related via the formulas $\tan\theta_W = g'/g$ and $g \cdot \sin\theta_W = e$, where θ_W is the Glashow–Weinberg angle.

and try to determine these amplitudes by means of a (perturbation) infinite series in the relevant interaction coupling constants. However, the appropriate series expansions are typically plagued by the appearance of highly divergent terms. In order to get finite results, physicists use *renormalisation* techniques which specify a way of 'regulating' the divergent integrals appearing in the problematic terms of the expansion and 'subtracting' the emergent infinities in a systematic way.

More precisely, the first step of renormalisation requires the positing of a specific 'regulator' which renders problematic integrals provisionally finite. The most intuitive method of doing this is by introducing an energy cut-off or, equivalently, a distance cut-off: This is in fact tantamount to (provisionally) ignoring effects that involve fields varying on distance scales shorter than the cut-off distance. The method, though relatively intuitive, is rarely used because, contrary to the popular method of *dimensional regularisation,* it violates Poincaré invariance, as well as any gauge invariance which a QFT may possess.

There are two ways of effecting the second step of the renormalisation procedure. One may adopt the way of *reparameterising* the coupling constants. The original, intuitive idea that the mass(es) and coupling(s) which appear in the original Lagrangian represent finite, measureable quantities is dropped; what is now assumed is rather that they are in fact *infinite* quantities. These quantities are taken to correspond to properties of 'bare' particles—that is, particles considered in the absence of any interaction (including self-interaction). But since such interactions are *always* present, 'bare' couplings are non-empirically determined entities amenable to arbitrary redefinition. By taking these quantities to be functions of the regulator, it can be shown that such redefinition can be done in such a way that the divergences of the integrals are *absorbed* into the redefined parameters. Thus, expressions for the amplitudes are produced which are both finite and independent of the regulator.

According to the second, more commonly used, approach, we do not interpret the constants that feature in the original Lagrangian as infinite quantities. We rather think that the original Lagrangian is *inadequate* to represent the correct structure of the interactions in question. To make it adequate, we can add regulator-dependent 'counterterms' the forms of which are carefully chosen so that the divergences, and also the regulator-dependence, disappear in the infinite limit. These new terms involve new constants with values which cannot be deduced from the theory itself but must be determined by experiment.

It is obvious that during the whole renormalisation process physicists make some arbitrary choices in order to define the renormalised quantities. Physics, however, cannot be sensitive to these arbitrary choices. Hence, the S-matrix (i.e., the theoretical tool which represents all the observable quantities associated with the theory) must be *invariant* under the transformations of possible reparameterisations. Such transformations form the *renormalisation group.* Demanding this invariance under reparameterisations gives rise to specific equations that inter alia describe how coupling constants vary as the renormalisation scale and the interaction energy change.

Now, according to its initial interpretation, the above-described technique has the desired result only in a certain limited class of physical theories which are called *renormalisable.* From the perspective of this construal, the feature of renormalisability introduces a selection criterion for physical theories. Only those theories satisfying it have any physical sense, and it is a very fortunate case indeed that almost all theories of fundamental interactions are renormalisable.

The basic idea behind the most recent approach to renormalisation is the following. We begin again with a QFT described by a Lagrangian with a set of 'bare' coupling constants. Starting off with these bare constants (the list of which we may think of as a vector in an appropriate abstract space), we can compute the 'physical' coupling constants at any energy scale. To do calculations with the theory we also need to ignore effects occurring at length scales smaller than the cut-off distance. If we know the vector of 'bare' coupling constants and the cut-off, we can compute the vector of 'physical' coupling constants at any distance scale. It is important to notice that the above procedure can take place whether or not our field theory is renormalisable in the older sense. In the case of a non-renormalisable theory, the physical coupling constants (at a distance scale different from the cut-off distance) in the non-renormalisable terms get smaller and smaller, approaching zero. At large distances, therefore, non-renormalisable interactions become irrelevant for the calculations. From that perspective, we may *explain* why the theories constituting the

Standard Model are renormalisable. For regardless of what's going on at very short distances, non-renormalisable interactions are bound to be very weak at large distances.

Having presented (in rather brief and superficial terms indeed!) the basic tenets of the renormalisation techniques, let us return to the step in which the Lagrangian of a QFT is 'filled' with counterterms in order to yield finite results.[4] Our aim in doing so is to stress a point that is crucial for the argument that follows. Suppose we deem the bare coupling constant of some interaction to be zero. That is, suppose we think that in our fundamental theory a certain interaction would not happen. Given that, we think we are justified to omit any term representing the interaction in question from our theory (for, whatever the form of the interaction, the relevant term would be zero since the associated coupling constant is zero). According to what has already been noted, however, this thought might be false. Despite the zero value of the *bare* coupling constant, the interaction might have a non-zero *physical* coupling constant at some energy scale. For instance, in the case of mass (or charge) of a given field (or of a particle interpreted as an excitation of the field) we may be falsely tempted to omit terms for all possible interactions of the given field (particle) which depend on that mass or charge. Since, however, the associated physical coupling constant might be non-zero at some higher energy scale, we should admit that the field (or particle) *acquires* a mass/charge at this energy scale. Crucially, for our purposes, this seems to indicate that, contrary to our intuitions 'flowing' mostly from a highly idealised physical description of the situation, all fundamental properties related to fundamental interactions are not 'given' *prior* to them but rather 'acquired' from them (see Butterfield and Bouatta (2015)). The following argument builds on that remark by associating a metaphysical interpretation of that priority. What I intend to show is that, given my preferred definition for the dispositionality of a fundamental property, the above-mentioned priority, properly construed, leads to the following conclusion: The most successful theoretical framework we currently have for fundamental interactions makes

[4] Divergences can *in principle* be cancelled even in the case of non-renormalisable theories. However, in that case we need to add to the Lagrangian *infinitely* many counterterms.

physical sense provided that scientists derive empirical results via a procedure which presupposes that fundamental properties related to fundamental physical interactions cannot be dispositional (and therefore, given that the dispositional/categorical distinction is exhaustive, they must be categorical).

Given the above background, my argument for the categoricality of fundamental properties is brief. We start recalling the @-criterion of dispositionality: The first-order state of affairs of an object instantiating a fundamental dispositional property is by itself (part of) a minimal truthmaker for specific (perhaps, world-relative) modal truths (expressed by specific non-trivial counterfactuals) which concern the ascription of specific (perhaps, world-relative) powers to the object. The first point we have to make is that fundamental properties of the actual world are various 'charges' intimately related to fundamental interactions, and so the modal truths that feature in the @-criterion are determined by physical science and concern these interactions. The second point is about the very notion of truthmaking which seems to induce some sort of *priority* relationship of an entity's existence over the truth of a truthbearer. In our case, the relevant entity is the first-order state of affairs of an object instantiating a fundamental dispositional property, while the relevant truthbearer is a proposition (or propositions) related to specific fundamental interactions associated with the property in question. The priority relation that concerns us here can be interpreted as a case of a metaphysical dependence-relation between particular sorts of facts. In other words, it seems plausible to think that there is a sort of relationship (not necessarily equivalence) between the truthmaking fact 'a makes proposition b true' and the fact 'a exists "grounds" the fact that b is true'. This notion of 'grounding' expresses the metaphysical dependence-relation between the existential fact and the fact that some proposition is true. Now, implementing the above relationship in the case of the truthmaking relation that features in the @-criterion we get:

@-GRND-criterion The fact that a first-order state of affairs of an object instantiating a fundamental dispositional property exists 'grounds' the fact that specific propositions about fundamental interactions are true.

The metaphysical dependence expressed by @-GRND-criterion is an *asymmetric* metaphysical relation which in the case at hand implies the following:

DISP-GRND It must *not* be the case that the fact that some propositions about interactions are true 'grounds' the fact that a state of affairs involving a dispositional property exists.

But it is this last claim that I think is refuted by the renormalisation practices of theoretical physicists. For what the phenomenon of 'acquiring' a charge due to the appearance of an interaction at an energy scale shows is that physicists treat fundamental interactions as *prior* to the possession of fundamental properties by elementary objects. Metaphysically construed, this priority (implicit in scientific practice) not only *allows* but also *dictates* that the following must hold: The fact that some propositions about interactions are true (namely, propositions about the interaction of an object with other objects due to the 'emergence' of a non-zero coupling constant at a higher energy scale) 'grounds' the fact that a state of affairs of the object in question instantiating a fundamental property-charge exists. The rejection of DISP-GRND in the actual world implies that fundamental charges *are not* dispositional properties; for if they were, DISP-GRND would be actually true.

3.2 Objections

There exist at least three objections to the argument from renormalisation. First, it might be objected that we should not draw metaphysical conclusions from QFT-based scientific practices, since the latter eventually breaks down at high-energy scales. Nevertheless, given the astonishing empirical success of QFTs, we have strong reasons to believe that any theoretical framework appropriate to describe high-energy phenomena will imply a QFT description at lower energies.

According to the second objection, we have not proved that *all* actual fundamental properties are categorical, since the argument from renormalisation does not work for the properties of the mediating bosons. As a first reaction, I may point out that there is a way to prove the categoricality of the *masses* of those bosons by appealing to the Higgs mechanism, an essential theoretical part of the Standard Model of elementary particle physics. The main theoretical role of the Higgs mechanism is to reconcile the experimentally determined short range of the weak force (a fact that requires that the bosons carrying the force have non-zero mass) with the renormalisability of the theory (which excludes the presence of mass terms in the relevant Lagrangian). Without going into any technical details, the implementation of the mechanism requires the existence of a field with non-zero vacuum expectation value—the Higgs field/particle—the *interactions* of which with the particles mediating the weak force and the elementary matter particles *give rise* to the masses of all particles.[5] Here, we find once again the *priority* of certain interactions over the properties, a fact we can use to argue (in a similar to my argument from renormalisation way) for the categoricality of all masses. Of course, my first response (being limited to the case of masses) does not meet the objection since there are *other* properties which the mediating bosons possess. For instance, particles W^+ and W^- that 'carry' the weak force possess electric charge and gluons that 'carry' the strong force between quarks possess colour. So here is my second response; since the debate between categoricalists and dispositionalists concerns properties of *matter*, the properties of the exchange bosons are irrelevant for the present discussion. This response presupposes, of course, an ontologically significant difference between matter and interactions; a difference which is arguably blurred in contemporary physics (especially in the context of theories based on supersymmetry). This last remark can be construed as showing that the dispositional vs categorical debate is ill defined in the modern scientific context. In that case, the crucial question is: Should

[5] According to the 'standard' textbook account of the Higgs mechanism, the generation of particle masses is achieved through the phenomenon of *spontaneous* breaking of a local gauge symmetry. In that phenomenon, the laws of motion of an elementary physical system retain their symmetry, which however is absent from the solutions of the associated equations. A number of philosophers have expressed qualms for such an interpretation on the grounds that local gauge symmetries do not correspond to any empirical features of the world. The case is controversial (Earman (2004); Friederich (2014); Lyre (2008); Lyre (2012); Smeenk (2006); Wuthrich (2012)); it seems however that this does not affect the theoretical role of the Higgs mechanism which, after all, can be implemented without assuming a gauge symmetry breaking (i.e., in a gauge invariant way) (see, for instance, Higgs (1966), Kibble (1967) and Struyve (2011)).

we dismiss the whole metaphysical issue or not? As far as I can see, the answer is no; yet, it is not an answer that can be generalised. Most probably, if we want to insist on the (at least partial) autonomy of metaphysical questions (a reasonable stance, I think, given scientific change) we need to be flexible and discuss them in a context sufficiently *close* to that in which they were originally defined. This suggestion, however, does not exclude the case where a metaphysical debate becomes *meaningless* in a specific scientific context. So, there must be a case-by-case examination of whether there is any room left for a discussion of a specific metaphysical issue. In the case at hand, there is room for a lot of fruitful metaphysical discussion of the associated metaphysical issue, at least within the context of the properties of matter as traditionally conceived.

The third objection is based on certain thoughts of Yuri Balashov (1999) concerning the genuine existence of zero-value properties. Indeed, if zero-value fundamental quantities do exist (on a par with the non-zero-value ones), one may object that elementary particles do not *acquire* a fundamental property through a specific interaction, but merely acquire a different *value* of a *possessed* property. In order to evaluate the strength of the objection, we have to take a closer look at Balashov's arguments in favour of the genuine existence of zero-value fundamental quantities.

Argument from Composition Balashov urges us to consider a particle which is a bound state of two other particles having non-zero (determinate) quantities falling under the determinable P and summing up to zero. He suggests that it is more reasonable to say that the original particle has zero value of P than to insist that it has no P at all. According to Balashov, P-hood cannot simply disappear when combined with another P-hood in a productive way. To put it informally, two or more P-hoods cannot result in complete P-lessness.

Argument from Parity Balashov argues that one may be justified to ascribe a zero-value determinate of the quantity P to a mereologically simple particle if the possession of the determinate of P by the latter can be related to a *relevant* common trait in the physical behaviour of this particle and another particle which is mereologically complex in the sense described in the Argument from Composition. As Balashov points out, the force of the Argument from Parity is proportional to the force of 'relevant' and can be increased by showing that the same *kind* of generic trait in behaviour is related to a *non-zero* value of P in other objects.

Argument from Unification According to this argument, we are justified to ascribe a zero-value determinate of the quantity P to a particle (rather than no P at all) if the latter exhibits, in common with particles possessing a non-zero-value determinate of P, a generic physical trait related to P-hood. Balashov claims that it would be unnatural to suppose that one particle behaves in a certain P-dependent way thanks to having P and another object behaves in the same way thanks to lacking P.

Argument from Disparity According to the last of Balashov's arguments, the claim that a particle possesses a zero-value determinate of a quantity P can be supported by showing that the particle in question differs from another one (which is unlikely to have anything to do with P-hood) in a physical trait known to relate to P-hood and its absence.

Having presented the four types of argument Balashov makes use of, it is time to see what their implementation can tell us about fundamental properties of elementary particles. Let us begin with the Argument from Composition. The existence of electrically neutral particles which are bound states of non-zero-charge components (e.g., neutron and its quarks), of bound states of quarks (baryons–mesons) which are colourless (though the quarks carry colour) and of spin-zero pseudo-scalar mesons (π^+, ρ^+) which are bound states of various pairs of quarks and anti-quarks, each having spin 1/2, guarantees the application of the argument in the case of all fundamental quantities. Nevertheless, I find this first argument inconclusive. First, it seems not to work in the case where fundamental properties are dispositional. Two determinates of the same determinable can be *finks* to one another. Consequently, and *pace* Balashov, their co-instantiation in a composite object may lead to their mutual *annihilation*. Second, I cannot see why the co-instantiation of *categorical* determinates cannot lead to mutual annihilations as well. Perhaps there are laws of nature to dictate that; or the application of fundamental symmetries might generate a *kind* of composites which must lack the relevant property.

Let me now turn to the Argument from Parity. The first thing we should point out is that it has no application in the case of colour; though there are strictly colourless non-composite particles (leptons), there are no non-composite particles with zero colour. It can apply, however, in the other cases. Consider, first, electric charge and compare a non-composite electrically neutral particle (neutrino) with a mereologically complex electrically neutral particle (neutron). They are perfectly on a par as far as their electromagnetic traits are concerned. So, given that we are justified to ascribe zero charge to the latter (argument from composition), the Argument from Parity suggests that the former (non-composite particle) has also zero charge.[6] It is obvious that the success of the Argument from Parity presupposes that we are already convinced (by the Argument from Composition) that composite particles may have zero-value determinate quantities. Otherwise, the argument has no force at all. Balashov himself presents a possible interpretation according to which the Argument from Composition and (consequently) the Argument from Parity fail. He envisages the possibility of considering the zero charge of a composite particle (such as neutron) as a supervenient property (supervening on non-zero charges) which under a certain construal is no addition of being. If that were true for composites, there would be no reason due to parity to postulate a zero charge for non-composite particles such as the Higgs boson or the neutrino. In fact we have reasons to reject such an attribution. In that (unfortunate for the arguments) case, the common traits of composite and non-composite particles are due to the *lack* of the determinable of charge.

Given the inconclusiveness of the Argument from Composition, the only way for the Argument from Parity to be useful is to be supported by another, more convincing, argument. This could prima facie be the Argument from Unification. Balashov points out that *each* generation of leptons and quarks includes a weak isospin doublet which crucially contains both a charged *and* a neutral particle. This fact suggests that the two particles must possess some common traits 'flowing' from the pos-

---

[6] Analogous remarks hold for spin. If we are justified to ascribe zero spin to composite pseudo-scalar mesons, the Argument from Parity indicates that we should do so in the case of non-composite spin-0 particles such as the Higgs boson. As Balashov points out, both obey Bose–Einstein statistics and have exactly one polarisation state. Hence, they do not differ in any spin-related physical traits.

session of a common determinable which, in this case, must be electric charge. Prima facie, it might be objected that electric charge is irrelevant since $SU(2)_L$ doublets are related to the *weak* interaction. But, Balashov argues, the electroweak unification in the context of the Standard Model removes the ground for such an objection. More precisely, to account for parity violation in weak interactions, the Standard Model introduces a new fundamental property, weak hypercharge Y. Parity violation is represented through two different kind of representations of SU(2): a *doublet* includes the neutrino and the left-handed electron and a *singlet* includes the right-handed electron (no right-handed neutrino exists). The members of the $SU(2)_L$ doublet are eigenstates of the component T_3 of weak isospin with eigenvalues differing by a unit of electric charge. Hence, Balashov concludes, contrary to the initial thought, electric charge *is* an essential property for members to be placed into $SU(2)_L$ doublets. I do not, however, find this argument convincing. Since the members of each fundamental $SU(2)_L$ doublet are eigenstates of weak isospin and, crucially, have the *same non-zero* value of weak hypercharge, it is much more plausible to assume that the common traits that those members must share to be able to belong to the same doublet are the specific determinate of Y and perhaps the determinable T_3. Another relatively significant shortcoming of the Argument from Unification is that it has no application in the case of colour. A (meta-)Argument from Unification would require of the original argument to be implemented with regard to all charge-like fundamental properties associated with fundamental forces.

In the case of colour, there is a clear *dynamical* difference between zero-colour hadrons (proton, neutron) and colourless particles (electrons, neutrinos) which seems to be able to ground an argument from *disparity*. Composite zero-colour particles *strongly* interact among themselves while colourless leptons do *not*. As Balashov argues, this fact supports the view that there is a clear sense of attributing a zero colour to the former and no colour to the latter. As it stands, this last argument from disparity is unconvincing. For, as far as I can see, what that fact really shows is an uncontroversial dynamical difference between coloured objects (quarks) which 'feel' the strong force and colourless objects (leptons) which do not. Given that the strong interaction between zero-colour composite particles is due to their non-zero-colour constituents, there is no cogent argument from disparity to defend the attribution of genuine zero-value determinates of colour to particles.

I conclude that there is no real threat from Balashov's arguments to my argument from renormalisation.[7]

[7] I do not discuss the case of mass since Balashov himself acknowledges that his arguments are far from conclusive in that case.

4

Categorical Monism and Quidditism

By far, the most common (among dispositional monists) strategy for arguing against the existence of purely categorical properties (and, *a fortiori*, for the falsity of Categorical Monism) is to appeal to its alleged problematic *modal* consequences. More precisely, it has been claimed that the 'nature' of categorical properties commits the categorical realist to accept counter-intuitive modal scenarios. In this chapter, I argue against this recurrent objection to Categorical Monism by challenging the necessity of this commitment.

4.1 Introduction

Consider a candidate for a natural fundamental property of our world, the electric charge. There are different *ways* in which a possible world represents various facts about *specific* actual fundamental properties, such as electric charge. In the philosophical jargon, the issue concerns how a world can represent properties *de re* (for instance, how a possible world w can represent the fact that charge *exists* and *does* so-and-so). According to the prima facie most convincing account, an actual

V. Livanios, *Science in Metaphysics*,
DOI 10.1007/978-3-319-41291-7_4

property has to exist *itself* in a possible world in order to be *de re* represented by it. If it does, we have a case of *de re* modal representation by *transworld identity*. There is, however, an alternative way of *de re* modal representation that is provided by *counterpart* theory. According to that approach, actual properties are not transworld entities; they exist only in the actual world. A possible world *de re* represents actual properties by having counterparts of them (which, according to some versions of counterpart theory, are similar to them in certain respects). Those counterparts are also world-bound entities existing only in the possible world under consideration.

Consider now two opposite charges attracting each other and ask whether there is a possible world in which the two charges repel each other. Or, alternatively and more generally, ask the following question:

Q_1 Is there a possible world in which the property of charge has a causal/nomic role different from the actual one?

As I understand it, Q_1 asks whether the *de re* modal representation of charge (and, in general, of any natural fundamental property) is determined or constrained in any way by a specific ontological factor, namely, its actual causal/nomic role. More precisely, we may first define a special way in which possible worlds can represent facts about fundamental properties. According to this way of modal representation (which we may call the *role-only* way), possible worlds can represent specific causal/nomic roles (profiles), but not specific properties that *fill* these roles. To be as clear as I can, I have to note that these causal/nomic roles can be construed as second-order properties of natural properties which can be defined using a formal procedure described by Hawthorne[1] (2001, 370). Here is a brief description: We start by constructing the Ramseyfied lawbook of the possible world by conjoining all the laws of the world in which the property in question exists, and then replace each property name (namely, each predicate that does not express a causal/nomic relation) by a distinct variable and prefix each variable with a quantifier. The open sentence that we get by dropping the existential quantifier prefixing the variable denoting

[1] Vallentyne (1998, 177–8) tentatively introduced this procedure in order to define the nomic role of any property, but rejected it outright for reasons briefly presented in his paper.

the property denotes the causal/nomic role of that property.[2] Armed with the above definition, we may rephrase Q_1 (in its generalised form concerning *any* natural fundamental property) as follows:

Q_2 Does the role-only way of modal representation of properties determine or constrain in any way the *de re* way or not?

There are two dominant views which provide divergent answers to Q_2. The first one, which I henceforth call the *Dispositionalist View* (DV, for short), intimately relates properties to their *actual* causal/nomic roles and further assumes that the latter *exclusively* ground or determine the *de re* modal representation of the former. More specifically, according to DV, a possible world w represents a specific actual property P as doing such-and-such iff, in w, a property that realises P's actual causal/nomic role is doing such-and-such. Adopting DV leads naturally to the claim that there are no possible worlds representing any property as having a causal/nomic profile differing from its actual one.

In opposition to DV, the second view, which I henceforth call the *Non-Dispositionalist View* (NDV, for short), allows different (from the actual causal/nomic roles) ontological factors to (at least) co-ground the *de re* modal representation of properties. In its extreme and most popular version, *Radical Non-Dispositionalist View* (RNDV, for short), it can be defined as the view according to which the actual causal/nomic roles of properties are *totally irrelevant* to their *de re* modal representation. And of course, in this case, the role-only way of modal representation of properties does not determine the *de re* way.[3] Adopting RNDV leads naturally to the claim that there are possible worlds which represent any property as having a causal/nomic profile differing from its actual one.

---

[2] Notice that the above procedure does not result in a complete Ramseyfication of the lawbook. This can be done only when *every* relational and non-relational predicate is replaced by a variable (including predicates denoting causal/nomic relations). This seems a more natural procedure for anyone who thinks that there is nothing special about causality/nomicity that vindicates the decision to be left unRamseyfied.

[3] According to the moderate versions of NDV, the former does not determine *exclusively* the latter.

4.2 Versions of RNDV

The above *negative* definition of RNDV excludes causal/nomic roles as relevant ontological factors for the *de re* representation of properties without, however, suggesting any other candidates to play the required modal role. Hence, a natural question that first comes to mind is:

Q_3 Besides the causal/nomic roles, which ontological factors *can* be relevant to the *de re* modal representation of properties?

Various versions of RNDV correspond to different ontological factors which may ground the *de re* modal representation of properties. In what follows, I'll provide a brief list of ontological options for RNDV-ists. The list is not meant to be exhaustive; yet, I think it achieves its main aim which is to show that RNDV is a much broader thesis than often assumed.

Option 1 A nominalistically (for the second-order features) inclined metaphysician may claim that there are *no* ontological features of first-order properties grounding their *de re* modal representation. It is the *properties themselves* (where the term 'themselves' is taken literally) that constitute the required ground.

This is the view that Dustin Locke (2012) embraces, calling it *Austere Quidditism*. It is instructive here to comment on the use of the term 'Quidditism'. Property theorists (following the lead of Robert Black (2000)) use the term to refer to what I call RNDV. I prefer, however, the latter term because the former gives the impression that RNDV has exclusively to do with the so-called *quiddities* (more on this in the sequel). But as the present list reveals, this is far from true.

Locke presents his view in terms of the *de re* modal representation of properties as follows:

> A possibility w represents a property P as being such and such way iff w is a possibility where P itself is that way. (2012, 357)

Austere Quidditism is clearly a case of *de re* modal representation of properties by transworld identity.[4] There are two points here worth mentioning because they often cause confusion. First, the fact that this version of RNDV excludes the appeal to counterpart theory of *de re* representation does not mean (as we shall see below) that other versions do the same. This remark is crucial because, as far as I know, it has been always assumed that RNDV and transworld identity of properties come as a package deal. A characteristic example of this attitude is Robert Black, who, in his thought-provoking work (2000), claims that Quidditism is so intimately related to the assumption of transworld identity of properties that one can reject Quidditism simply by holding that properties are world-bound entities. Second, the fact that Austere Quidditism excludes ontological *features* of properties from being grounds of the *de re* representation of the latter does not mean that it declines to answer Q_3. As I have already mentioned, it only insists that properties themselves are the required ontological ground. Black (like the austere quidditist) also seems to deny that Quidditism can accept any ontological features of properties to ground their transworld identity. Yet, he crucially thinks that this implies that Quidditism is tantamount to the acceptance of *primitive* transworld identity of fundamental properties, a fact that indicates the lack of *any* ontological ground of the *de re* modal representation of the latter.[5]

Option 2 The ground of the *de re* modal representation of first-order properties has nothing to do with second-order properties in general. It is rather a transworldly stable *property-substratum* (P-substratum) that is 'connected' with all (contingent and essential) second-order features of properties and grounds the *de re* modal representation of the latter.

[4] Locke thinks that this feature distinguishes Austere Quidditism from both DV and Quiddity Realism (a view which I shall present in the sequel). He claims that the aforementioned views are counterpart theories grounding the *de re* modal representation of properties on causal/nomic role-sharing and quiddity-sharing, respectively. But this is not true at least for the case of DV; dispositional essentialists define DV as an account concerning the transworld identity of properties. One and the same property is characterised by a specific causal/nomic role in all worlds in which it exists.

[5] '*Nothing* constitutes the fact that a certain quality playing a certain nomological role in that world is identical with a certain quality playing a different role in ours. They just are the same quality, and that's all that can be said'. (2000, 92, my emphasis)

This is a view that a metaphysician believing in the ontological complexity of first-order properties may naturally hold. P-substrata have no nature or properties (in a sense, of course, that needs to be clarified) and besides their role in the *de re* modal representation of properties they can also play at least two important metaphysical roles: First, they can be the *bearers* of second-order properties and the *unifiers* providing first-order properties with the required unity. In particular, the appeal to P-substrata can provide a brief solution to the problem of the unity of second-order properties[6] constituting a first-order property, provided that the relationship of P-substrata with second-order properties is a sui generis,[7] external relation which neither depends on nor is grounded in the P-substratum in any way. (According to this approach, P-substrata remain ontologically simple entities in spite of their 'connection' to properties; properties are not ontological constituents of P-substrata.) Second, they can be the metaphysical ground of the *intra-world* individuation of mereologically simple[8] fundamental properties. One may first claim that the individuation of causally/nomically indiscernible properties is due to the presence of a P-substratum in each of them. And then generalise her claim for any property as follows: Since any property *could have had* a causally/nomically indiscernible property, this possibility (which leads to the adoption of P-substrata) must be 'registered' with the ontological structure of the property in question. Therefore, every property must have a P-substratum as an ontological constituent. And this constituent, according to the suggested account, is also transwordly stable.

Option 3 The *de re* modal representation of first-order properties is grounded in some roles (second-order features) of properties. These roles, however, are not causal/nomic but *metaphysical*.

The motivation for this view is due to Mumford's (2004, 188) suggestion that there can be properties which are non-powers and bear only metaphysical relations to other properties. According to Mumford, the list of meta-

[6] Alternative accounts appeal either to a relation of compresence or to relations of ontological dependency between properties. For details about the problems that arise in the context of these accounts, see, for instance, Simons (1994) and Denkel (1997).

[7] Assuming that this relationship is an ordinary relation leads to a vicious regress.

[8] The individuation of complex properties can be grounded in the prior individuation of their mereologically simple property-parts. P-substrata can be construed as the *ultimate* individuators.

physical relations is the following: First, there are relations of *metaphysical necessity*, where the having of one property metaphysically makes necessary the having of another property; for example, being coloured makes necessary being extended. Second, there are relations of *metaphysical possibility*, where the having of one property makes metaphysically possible the having of another property; for example, having length makes possible being 10 cm long. And finally, there are relations of *metaphysical incompatibility*, where the having of one property is metaphysically incompatible with the having of another property; for example, having mass 50 g is incompatible with having mass 100 g.[9] Given the above-mentioned relations, one might envisage a version of RNDV in which the *de re* modal representation of properties is grounded in the pertinent metaphysical roles. The latter can be constructed via a method akin to the one suggested by Hawthorne; instead of starting from the Ramseyfied *lawbook* of nomic relations, we should start from the Ramseyfied book of metaphysical relations.

Option 4 Contemporary physics suggests that the intraworld identity of actual fundamental natural properties that dispositionalists themselves often invoke (such as mass, electric charge and spin) can be determined by a procedure *independent of* their causal/nomic roles. In particular, fundamental properties can be identified as invariants under the action of transformations associated with fundamental symmetries (external or internal).[10] Given that the intraworld identity of these fundamental physical properties can be provided in the actual world via symmetry considerations, there is nothing, I suggest, to prevent one from claiming that the second-order features 'being invariant under the action of the Poincaré group of symmetry transformations' and 'being invariant under the action of the U(1) group of symmetry transformations' (the *invariance-features* henceforth) can *also* ground the *de re* modal representation of those properties.[11]

[9] Mumford (2004, 177) also introduces relations of *dispositional* necessity and possibility. But only the metaphysical relations mentioned above can be considered as a ground for a version of RNDV.

[10] See also Psillos (2006) and Livanios (2010).

[11] From the perspective of the transworld identity theorist, invariance-features, which are the grounds for the intraworld identity of mass and spin and charge, can be construed as *the* grounds for their transworld identity too.

(It is important to stress at this point the difference between what I propose here and the remarks made in Sect. 2.2.1. Though both suggestions are inspired by the same symmetry considerations, they concern different issues. The topic of the Sect. 2.2.1 concerns the possible *predicates* that we may use to describe and, at least partially, identify fundamental properties, whereas the issue under consideration here is the available *ontological* factors that may ground the *de re* modal representation of properties.)

The obvious objection to the preceding remarks is that invariance-features are *not* independent of the causal/nomic roles of fundamental properties. First, it might be objected that it is not clear that the suggested features are completely independent of the *causal* roles of properties. After all, in a certain sense, invariance under the action of symmetry transformations involves the comparison of specific magnitudes before and after certain operations. The values of these magnitudes ultimately come from the readings of scientific instruments. On this view, when we say that a quantity is invariant, what we actually mean is that a particle (perhaps an elementary one) interacts with an instrument in the same way in different (yet related by symmetry transformations) circumstances. In other words, certain symmetry-related circumstances do not affect specific causal roles of the measured property of the particle. We have, therefore, an *implicit* introduction of these causal roles in the suggested procedure of identification which turns it into a kind much like the one accepted by dispositionalists. Nevertheless, this objection relies on an 'operationalistic' approach to the issue of invariants. According to this view, in order to find whether a quantity is invariant or not, we have to measure its value in different circumstances and compare the results. Yet, what I have described in Sect. 2.2.1 is a purely *conceptual* way of finding the invariants under certain transformations, given the group structure emerging from their repeated application. Hence, insofar as we do not have to measure anything in order to find the invariants (in the case of mass, charge, etc.), it is reasonable to claim that the suggested procedure of identification and the invariance-features grounding the *de re* modal representation of the fundamental properties in question are both independent of their causal roles.

According to a second objection, invariance-features are *metaphysically related* to the *nomic* roles of the properties they characterise. Consider, for instance, the property of mass and the invariance-feature 'being invariant under the action of the Poincaré group of symmetry transformations'.

Assume now that (a) the symmetry involved in the definition of the invariance-feature is a *property* of each member of the web of nomic relations which determine/constitute the nomic role of the property at issue; and (b) properties are modes of existence of their bearers and, so, are metaphysically dependent on them. Given those assumptions, one might claim that the invariance-feature is in an *implicit* manner metaphysically related to the nomic role of mass, because it involves a kind of symmetry which is metaphysically dependent on the nomic relations determining this role.

It is certainly true that fundamental symmetries (such as the Poincaré symmetry) can be construed as properties of nomic relations. Hence, if we also grant the objector her conception of properties as metaphysically dependent entities, the property 'being invariant under the action of the Poincaré group of symmetry transformations' is metaphysically related to the nomic role of mass. But, in order for the objector's argument to work, this invariance-property must characterise the nomic relations and not the property of mass: it must be the well-known property of form-invariance of nomic relations under the action of the particular symmetry. To conclude, though both this invariance-property and the above-mentioned invariance-feature involve the Poincaré symmetry, they are nevertheless different properties. The former is a property of nomic relations, while the latter is a property of mass. Hence, even if we grant that the objector's argument is sound, this does not prove that the second-order property, which I propose as a possible ground of the *de re* modal representation of mass (i.e., the invariance-feature), is metaphysically related in any way to the nomic role of the latter. (Incidentally, we can respect the connection between nomic relations and symmetries without insisting that the latter are metaphysically dependent on the former. We may claim that symmetries constrain nomic relations and are metaphysically independent in the sense that they can still exist even in possible worlds with nomic relations different from those in the actual one.)

The objector may challenge this conclusion by considering the case of electric charge. As I have already remarked, electric charge can be identified with the *conserved* quantity which results after the application of Noether's theorem in the case of the action (on the Lagrangian of a complex scalar field) of the internal continuous symmetry transformations U(1). The objector may start from that fact and generalise my suggestion by defining invariance-features corresponding to other fundamental symmetries besides the Poincaré symmetry. And she may further claim that there is no barrier to regarding the invariance-feature under the internal symmetry U(1), which grounds the intraworld identity of charge, as the ground of its *de re* modal representation as well. The objector may find the case of charge more congenial to her view, because in this case it seems that invariance under the action of a symmetry is intimately (and under a certain interpretation, also metaphysically) related to a particular feature (conservation) which *can* be related in turn to the nomic roles of charge. For, obviously, the conservation *law* of charge is a law involving charge and, under a certain conception, is among the factors constituting the nomic role of charge. Yet, once again, the real work is not done by the invariance-feature as I defined it. Since the invariance-property under the action of U(1) (which yields the conservation of charge law) concerns the Lagrangian of physical systems and is clearly distinct from the invariance-feature characterising properties, the new argument of the objector (like the previous one) fails to prove the existence of any relationship between the nomic roles of charge and the invariance-feature under U(1).

Option 5 According to the final option available to the RNDV-ist, the sole ground of the *de re* modal representation of any first-order natural property is its *quiddity*.

I call this view *Quiddity Realism* because it presupposes the genuine existence of quiddities. For my purposes it is the most important version of RNDV because, inter alia, most of property theorists believe that it is the *only* available version. But what is a quiddity? As Dustin Locke (2012) remarks, there is an ambiguity in the term 'quiddity'. According to one use of the term, quiddities (*quidditates* in scholastic tradition) are just first-order properties that describe the essential natures of concrete individuals. A quiddity of a property is nothing other than the property itself. A prominent philosopher who has identified quiddities with first-order properties is David Armstrong. The identification, however, is made only between quiddities and a certain kind of properties, the categorical ones. In his (1997), Armstrong identifies natural categorical properties with *thin* quiddities which, within each adicity class of properties, differ merely numerically.[12] While in his earlier work (1989), Armstrong had endorsed a different, *thick* sense of quiddities which differ from each other by their *nature*. Now, if we adopt this view (i.e., quiddities as first-order properties), then Option 5 is *almost* identified with Option 1. I say 'almost' because, unlike Option 1, Option 5 does not rule out the possibility of fundamental properties *de re* modally represented via their counterparts.

According to a different interpretation of the term 'quiddity', quiddities of properties are distinct (though non-separable) from properties themselves. The term 'quiddity' refers to an intrinsic *aspect* of a property, a second-order property. How can we construe in this case the intrinsic feature of properties that we call quiddity? As a first attempt, we can provide a *linguistic* account of such a second-order property:

Df$_1$ A quiddity is a second-order property designated by an expression of the form 'the property of being identical with P', where 'P' is a predicate referring to a first-order property P.

But obviously, for the ontological purposes of this book, we need a non-linguistic definition of quiddity:

Df$_2$ Q is a quiddity = (P) (Q is the property of being identical with P), where P is a first- order property.[13]

Quiddity seems to be an intrinsic, albeit relational, second-order property. The reason is that it incorporates, in a sense, the *relation* of identity. Nevertheless, it does not coincide with the relation of identity of a specific property with itself. It would be a category mistake to identify quiddity (a monadic property) with identity (a relation). The same can be argued for the case where someone claims that quiddity is a kind of 'unsaturated' two-place relation with one of its places filled

[12] Susan Schneider (2001) highlights some difficulties that emerge in the context of Armstrong's own metaphysics as a consequence of the identification of properties with 'thin' quiddities.

[13] It might be objected that quiddity, *qua* entity not able for multiple exemplification, is not a property. But I think, following Rosenkrantz (1993), that the capability of exemplification by different entities is not a logically necessary condition of propertyhood (but the having the feature 'being exemplifiable' is a logically necessary condition).

in by a specific property (see Rosenkrantz (1993, 107–114) for *mutatis mutandis* similar remarks for the case of haecceity of concrete particulars). Depending on one's preferred notion of non-qualitativeness, one might characterise quiddity as qualitative or non-qualitative property. For some philosophers (see, for instance, Bird (2007, 71)) any property incorporating in *any* sense the relation of identity is non-qualitative. (This belief, I think, is based on an analogy with the haecceity of concrete particulars.) For them, therefore, quiddity (as defined above) is a non-qualitative second-order feature. Yet, in another sense, this is not true. According to a more *restricted* account, a property is qualitative when it does not pertain to a specific *concrete* entity in a certain (intimate) way. Quiddity (*qua* second-order property) does not pertain to any specific *concrete* particular, and so, in this sense, is a qualitative property (Rosenkrantz 1993, 6).

In presenting available versions of RNDV, I've not taken sides in the debate between transworld identity theory and counterpart theory. The reason is that I think the core idea of *each* version can be articulated within the context of both theories of *de re* modal representation. This is obviously so for Options 3, 4 and 5, but it is also true even for Options 1 and 2 which commit to the transworld identity of first-order properties and the P-substratum, respectively. As I've already remarked, the core tenet of Austere Quidditism (namely, that the ground of *de re* modal representation of properties is the properties themselves) holds in one version of Quiddity Realism which (crucially) does *not* commit to the transworld identity of properties. Furthermore, the P-substratum of Option 2 need not *literally* exist in all worlds in which a first-order property is *de re* modally represented. In both cases, we may invoke *counterparts* of the entities involved and the appropriate counterpart relations between them in order to advance counterpart-theoretic versions of Options 1 and 2. In the sequel, while defending the possibility of RNDV against a major objection, I'll present details of how this can be done. I'll also show there that, though both contexts are prima facie available to the RNDV-ist, the counterpart-theoretic context is the only one which can provide a way out of the major difficulty besetting her view.

4.3 Defending RNDV

In this section, I concentrate on the major (in my view) difficulty besetting all forms of RNDV, the so-called *Permutation Difficulty*. First, I state the problem in Sect. 4.3.1 and then (in Sects. 4.3.2 and 4.3.3) I describe the steps that a proponent of RNDV may take to respond to it. Those steps include the adoption (a) of the world-bound existence of fundamental natural properties, and (b) of a counterpart-theoretic framework for their *de re* modal representation. Due to the controversial nature of both views, I found it necessary to defend at least their possibility (as alternative metaphysical frameworks) by addressing some relevant objections. Finally, in Sect. 4.3.4, I present two versions of the property-counterpart framework which are consistent with RNDV and argue that one of them can provide an adequate solution to the permutation problem. The conclusion is that, despite claims to the contrary, the permutation difficulty poses no threat to RNDV. So, even if there is an *exclusive* relationship between categorical features and RNDV (an assumption that I dispute), the permutation difficulty cannot be the ground for a decisive objection to Categorical Monism.

4.3.1 The Permutation Difficulty

A number of opponents of RNDV present an argument against it which I shall henceforth call the Permutation Argument[14]; some of them believe (as I also do) that it is the major argument in the debate between DV and RNDV. We may find versions of it in Bird (2007, 73–74), Mumford (2004, 151–152), Kistler (2002) and, for discussion about its epistemo-

[14] Lewis (2009, 208) also calls his first epistemological argument for Humility 'The Permutation Argument'. According to Lewis, the relevant permutations concern the actual referents of the theoretical terms (T-terms) of the Ramsey sentence of the true and complete final theory. His sceptical conclusion is that those permutations generate (due to RNDV) possible realisations of the final theory distinct from the actual ones, and no possible observation can tell us which one is the (unique) actual one. In the remainder of Sect. 4.3, I'll continue using the term 'Permutation Argument' for the argument as I describe it (namely, as concerning our modal intuitions and not our epistemic predicament).

logical consequences, Shoemaker (1980), Lewis (2009). According to the defenders of the argument, adoption of RNDV unavoidably leads to the view that possible worlds differing only in the specific natural fundamental properties which fill each causal/nomic role in those worlds (but not in any other respect) are distinct. (Typically these distinct worlds are 'generated' by permutations[15] of the properties among the causal/nomic roles they occupy, a fact which justifies the choice of name for the argument.) But, as the argument goes, our modal intuitions compel us to reject the existence of these distinct worlds. For, it seems difficult to understand how fundamental natural properties can swap causal/nomic roles and remain the same. It seems difficult, for instance, to understand how, in another possible world, electric charge can play the actual role of rest mass and vice versa. We can also show, by presenting the Permutation Argument as an analogue of Chisholm's (1967) argument against haecceitism, how the advocates of DV may exploit the argument to defend their view (indeed, this is what Alexander Bird (2007, 73) does). We may imagine 'small' changes in the causal/nomic roles of two fundamental natural properties in a sequence of distinct possible worlds. Assuming that those 'small' changes do not alter the identities of the properties under consideration, we may ultimately reach a possible world in which the two properties have swapped roles. But here our modal intuitions protest that there should not be such a world which differs from the actual one. So, if we take our modal intuitions seriously, there are at least two available options for us. We have to either abandon the hypothesis of the transworld identity of fundamental natural properties (therefore, it is not the case that the *same* natural property plays different roles in distinct worlds) or reject the view that all features of properties (including their causal/nomic roles) are metaphysically contingent (therefore, it is not the case that *any* change of them leaves the identity of properties intact). Bird goes for the second option, arguing that the causal/nomic role of any fundamental natural property (a role which, for him, is completely determined by the property's powers) is *essential* to it. In what follows, I shall examine, on a par with Chisholm's choice

[15] Note that this procedure does not require the permutation of properties among the places defined by the pattern of their instantiation.

in the original argument, whether a convincing solution to the problem may be found by following the first option. But before embarking on this task, one point must be made. As I have already noted, it seems that, if role-swapping is possible, RNDV runs contrary to our modal intuitions. Hence, an easy way out for the advocate of RNDV is simply to reject the authority of these intuitions on the matter under dispute. She may claim that the permutation problem springs from intuitions which *presuppose* that causal/nomic roles at least co-ground the *de re* modal representation of fundamental natural properties. And she may go on to say that this is not something that should surprise us; plausibly, modal intuitions emerge from our acquaintance with the world, and in the case of natural properties have to do with their causal/nomic roles. Hence it is not absurd that the possibility of swapping invites a kind of uneasiness. But if we acknowledge the fact that other ontological factors can be the exclusive grounds of the *de re* modal representation of fundamental natural properties, then we should not think it strange if properties have different causal/nomic roles in different possible worlds. Besides, as a number of defenders of DV (especially, dispositional essentialists) advise us (see, for instance, the case of the metaphysical contingency of laws of nature), we should not always *trust* our modal intuitions. Perhaps, rejecting the authority of modal intuitions provides a defensible way to avoid the permutation problem.[16] But it is not very appealing as it stands. In my view, the uneasiness springing from our modal intuitions should urge us to search for an appropriate *metaphysical* framework in the context of which a defensible answer can be given. Only then, I claim, the defender of RNDV could have good reasons not to trust the dispositionalist modal intuitions. So, in what follows, I present the two steps an RNDV-ist may take in order to ground an appropriate metaphysical framework within which a solution to the permutation problem can be provided.

[16] In his contribution to a recent volume (2009), Armstrong suggests that *in this case* we should not take our modal intuitions seriously because they rely on *mere* possibilities. He justifies his proposal on the basis of his 'chauvinism' for the actual world and declares that metaphysicians should not put much weight to arguments (such as the Permutation Argument) which rely on mere possibilities.

4.3.2 Step 1: Rejecting the Transworld Existence of Fundamental Natural Properties

The force of the modal intuitions behind the permutation problem is intimately related to the hypothesis about the transworld identity of natural properties swapping causal/nomic roles. As I have already remarked, it is difficult to understand how the very same fundamental natural properties can swap their actual roles in another possible world. So, if the defender of RNDV doesn't want to follow the essentialist course (recall the discussion on Chisholm's version of the Permutation Argument), the first step she has to take to relieve the tension is to reject the transworld existence of properties. *Pace* Lewis (1986, 205), she may claim that at least first-order fundamental natural properties cannot exist in more than one possible world. It is beyond the scope of this work to attempt a detailed defence of the thesis of world-boundedness of natural properties or examine its merits relative to the transworld-existence view. Rather, my modest goal here is to show that it is not an untenable thesis and, therefore, can be used in order to set up a metaphysical framework for a solution to the permutation problem. To illustrate this, in what follows, I examine two objections aiming at proving that natural properties are *necessarily* transworld entities and show that both fail.

World-boundedness of fundamental natural properties is not a very popular view. It is an undeniable fact that most philosophers engaging in the debate between DV and RNDV tend to assume the transworld existence of natural properties.[17] My best bet is that this tendency has mainly to do with their preferred view about another issue in the metaphysics of properties, namely, the universals/tropes debate. Let me explain why; most of the participants in the debate between DV and RNDV *assume* that fundamental natural properties are universals which, by definition, are strictly identical in all their instances. But, one may claim, the only way to secure that natural properties are strictly identical in all their actual

[17] Even David Lewis, who adopts the world-boundedness of individuals, assumes that natural properties are transworld entities. Lewis (1986, 205) claims that, due to the principle of recombination (i.e., a duplicate of anything can co-exist, under some restrictions, with the duplicate of anything else), there is a qualitative duplication between his concrete possible worlds. Since universals are supposed to recur whenever there is duplication, they are common parts of many worlds.

and possible instances is to assume that actual natural properties themselves live (at least) in all possible worlds in which they have instances. Hence, a first worry about the hypothesis of world-boundedness is that Universalism (namely, the view that all fundamental natural properties are universals) *requires* the transworld existence of properties. A quick reaction would be that there is no cogent reason why property theorists should be universalists. Trope theory is surely a viable account of the nature of natural properties and has recently gained popularity. But I do not need to appeal to the alleged truth of trope theory to meet the objection. Besides, as Tugby (2012) argues, it may be the case that embracing Universalism is the most plausible way available to the (pan)dispositionalist to meet the challenge of providing a theory about dispositions able to account for directedness in an intelligible way, while at the same time preserving the fact that many dispositions may be instantiated intrinsically and may also exist unmanifested (2012, 170). Hence, in order not to beg the question against a defensible metaphysical account of the nature of properties, I prefer to address the difficulty by arguing that the above conclusion (namely, that universals *must* be transworld entities) does not *necessarily* follow from the definition of a universal. A universal is an entity for which it is *possible* to have multiple instances. This *de re* possibility can be understood either according to transworld identity theory or according to counterpart theory. Universalists are free to adopt the view that actual universals are world-bound and have counterparts in other possible worlds. In this case, when universalists say that an actual universal is possibly multi-instantiated, they mean that at least one counterpart of the universal has multiple instantiations in the possible world in which it exists. Furthermore, though it is (by definition) true that any universal is strictly identical in all its instances, nothing can prevent the universalist to insist that this is not a strict cross-world identity, but a world-bound one. Each universal is strictly identical in all its instances existing in the world in which it exists. And each counterpart of the universal is also strictly identical in all its instances existing in the world in which it exists. It seems, therefore, that universalists, *qua universalists,* have no cogent reasons to reject the world-boundedness of natural properties.

The second worry I want to discuss is based on an analogy between temporal and modal repeatability. Schaffer (2005, 15) makes use of that analogy to argue that it only makes sense to regard individuals as world-bound (non-repeatable across possible worlds) and properties as transworld (repeatable across worlds) entities. But, even if we grant Schaffer that temporal repeatability implies modal repeatability, his conclusion depends on the assumption that individuals are *actually* not temporally repeatable, while properties (property types) are. This claim, however, holds only under certain other metaphysical assumptions which we may possibly reject. To illustrate this, consider first the claim that individuals are not temporally repeatable. The truth or falsity of this claim depends on the preferred view about the persistence of individuals.[18] If we think that individuals endure, then by definition they are repeatable in time and the claim is false. If, on the other hand, we think that they perdure, it is only when we think of their *momentary* temporal parts as the only genuine existing individuals that it makes sense to talk about temporally non-repeatable individuals. Consider now the second claim, the one about the temporal repeatability of properties. It is far from universally accepted that properties are temporally repeatable. A growing number of metaphysicians believe that the fundamental constituents of the actual world are particularised properties, that is, tropes. And one way to understand their particularity is to construe them as spatially and temporally non-repeatable. The objector may of course insist that her argument appeals to property *types* which, within the trope-theoretic context, can be plausibly construed as *transworld* classes of exactly resembling tropes. This assumption, however, does not imply that property types are repeatable across worlds; it only implies that property types have *parts* which inhabit different possible worlds. Modal repeatability, in the sense that one and the same property exists *in its entirety* in different possible worlds, is not ensured either way. The upshot, then, of the above remarks is that it is not necessarily the case that individuals are not temporally repeatable, while

[18] Roughly, *persistence* occurs when something exists at more than one time. According to *endurantism*, objects persist through time by being wholly present at a succession of moments, whereas, according to *perdurantism*, objects persist through time by having temporal parts (just as they extend through space by having spatial parts). For details, see Hawley (2001) and Haslanger and Kurtz (2006).

properties (property types) are; and consequently (following Schaffer's line of argument) it is not necessarily the case that natural properties should be regarded as transworld entities.

To my knowledge, there is no other argument against the possibility of world-bound fundamental properties. Hence, my arguments so far show that (at least) the defender of RNDV can use the hypothesis of world-boundedness in order to provide a solution to the permutation problem. There is a final worry, however, that must be addressed before we proceed. It seems that the assumption that all fundamental natural properties are world-bound entities (and so that the actual properties live only in the actual world) solves the permutation problem only by making their causal/nomic roles *essential* to them. But this is exactly what dispositional essentialists assume when they claim that causal/nomic roles constitute the individual essences of properties (putting it differently, for dispositional essentialists the alleged essential character of roles is the main reason for treating them as the sole ground of the *de re* modal representation of fundamental natural properties). Hence, it might be objected, the suggested step for the solution to the permutation problem undermines the very thesis of RNDV. This is a prima facie serious objection which, however, can be easily met. First, RNDV-ists may point out that the assumption of world-boundedness of natural properties may at best introduce a kind of metaphysical necessity[19] for their causal/nomic roles which, however, does not entail that the latter are *essential* to the former as well. A metaphysically necessary feature does not have to be an essential one (see Fine (1994) for relevant examples). Furthermore, it is possible that the metaphysical necessity of causal/nomic roles is due to some kind of primitive metaphysical necessity of laws of nature and has nothing to do with the transworld identity of the property in question.[20] Therefore, it is not necessarily true that the assumption of world-

[19] A metaphysically necessary feature of an entity is one that the latter has in each world in which it exists. This condition seems to be trivially satisfied in the case of world-bound entities and all of their features. But on closer inspection this is not true, because it presupposes a theory of *de re* modal representation (transworld identity theory) which, by definition, we cannot invoke in the case at issue. See also the second response to the objection.

[20] Fales (1993) tentatively suggests such a view when he claims that the necessity characterising causal/nomic relations is primitive and sui generis.

boundedness of properties turns RNDV into a version of DV. Second, and more importantly, RNDV-ists may argue that world-boundedness does not even imply any kind of metaphysical necessity of causal/nomic roles. The modal profile of any world-bound entity must be assessed in the context of the counterpart theory for *de re* modal representation. In that context, a metaphysically necessary feature is one shared by all counterparts of the entity. But this cannot be true *just* because the entity under consideration is world-bound.

4.3.3 Step 2: Adopting the Counterpart Framework

If we grant that the remarks of the above sub-section are persuasive, the assumption about the world-boundedness of fundamental natural properties seems to dissolve the permutation problem. It renders the scenario of role-swapping literally *impossible*; there is no possible world in which two actual fundamental natural properties swap their roles, simply because there is no world distinct from the actual one in which the properties under consideration exist. Furthermore, as I have remarked, the associated view does not coincide with the dispositionalist approach. Yet, despite its initial attractiveness, the world-boundedness hypothesis by itself is hardly adequate for a solution to the permutation difficulty. The main reason is that it dissolves the problem only by forbidding any kind of modal judgement about the possible behaviour of fundamental natural properties and, consequently, by totally ignoring the intuitions behind the role-swapping scenario. To put it differently, we need a theory of the *de re* modal representation of fundamental natural properties to provide the means required to make sense of the permutation problem; and simply suggesting that properties are world-bound does not offer one.

The assumption of the world-boundedness of fundamental natural properties excludes the appeal to the transworld identity approach and puts the discussion in the context of counterpart theory. Exploiting the counterpart-theoretic framework is not a novel approach to the *de re* modal representation of properties (see, for instance, Heller's (1998) suggestion about the existence of property-counterparts in the context of his linguistic ersatzism); yet, just like the world-boundedness hypothesis, it is not a popular view among philosophers participating in the debate between DV and RNDV. A widely held assumption is that RNDV is consistent only with the literal transworld identity of natural properties. The core intuition behind this stance is that RNDV requires that there be worlds which are such that a property filling a certain causal/nomic role in some of them fills a different role in others. And this can only hold if the property *itself* exists in all worlds under consideration. But that is too quick. Unless we give some good reasons against the implementation of counterpart theory in the case of properties, RNDV-ists may claim that a fundamental natural property can be *de re* represented in a possible world by having a counterpart in that world. Of course, the plausibility of such a claim depends inter alia on whether RNDV-ists can address the general objections against counterpart theory of natural properties. To this end, in what follows, I examine two such objections. According to the most often discussed objection, adoption of the counterpart-theoretic framework for natural properties introduces an *unnecessary* complication into the metaphysics of modality with no obvious motivation. The situation is contrasted to Lewis' counterpart theory. Lewis (1986, 201) famously presents as his main motivation for invoking counterpart theory for individuals the dissolution of the problem of *accidental intrinsics*. The assumption of the transworld existence of individuals implies the absurdity that one and the same individual has *and* does not have any of its accidental intrinsic features. But, according to the objection, nothing analogous holds in the case of natural properties. Arguably, the only intrinsic features that any alleged transworld fundamental natural property could have are its nature (whatever that may be)[21] and/or a (distinct from its nature) non-qualitative individuating factor. However, the objection goes on, both of them cannot be accidental features of the property.[22] Hence, given that there is no problem of accidental intrinsics for proper-

[21] Here the term 'nature' is not confined to the dispositional/categorical character of properties. Lewis (1986, 205) discusses the issue in the case of properties construed as universals and presents the ontological simplicity (or its opposite) as an intrinsic feature of a universal that most probably belongs to its nature. On a par with my comments, he claims that this feature seems to be an essential one.

[22] Of course, according to neo-Humeans, the causal/nomic role of any natural property is an accidental feature varying across worlds. Yet, it is also an *extrinsic* feature, because it is 'related' to external (for the neo-Humeans) relations between the property in question and other properties. As

ties, no motivation for appealing to counterpart theory can originate in it.[23] This remark, however, does not imply (*pace* Schaffer (2005, 15)) that the motivation for invoking counterpart theory is missing. Due to the flexibility of the counterpart-theoretic framework, both opponents and proponents of Dispositional Essentialism (and property-identity theorists as well) may appeal to it for their own ends (epistemic and metaphysical). First, property-counterpart theory can ground anti-essentialism for natural properties, because what is essential to a specific property may be different in different contexts (depending on which respects of similarity are important to us in each context). This feature of counterpart theory may be exploited by opponents of Dispositional Essentialism who seek a reason to deny that natural properties have absolutely determined essences related to their actual causal/nomic roles (see Heller (1998, 309–311)). Moreover, a proponent of RNDV may appeal to counterpart theory both for individuals and for properties in order to block the sceptical argument for Humility that Lewis (2009) presents.[24] Second, a proponent of (moderate) DV[25] may use the counterpart-theoretic framework to accommodate the possibility of distinct natural properties having the same causal/nomic role as an actual one. If the transworld identity-theoretic framework is the proper one for natural properties, then moderate DV-ists have a difficulty to account for this possibility because distinct properties cannot be identical to the one actual property having the aforementioned role. But given that counterpart relations need not be injective, distinct natural properties can be counterparts of a single actual natural property (see Ball (2011, 9)). Third, a property-identity theorist may utilise the means of counterpart theory in order to offer explanation of how

Lewis (1986, 201) persuasively argues, however, extrinsic second-order features do not pose any problem for the transworld identity theory of *de re* modal representation.

[23] Derek Ball (2011) claims that we may utilise Egan's (2004) remarks to create a problem from accidental intrinsics for properties construed as universals. The alleged accidental intrinsic feature of a natural property is the (second-order) property of being instantiated. Yet, this suggestion is controversial; first, because it is not at all clear why we can think of this feature as *intrinsic* (can a lone natural property have it?) and, second, because according to some theories of universals (such as Armstrong's) being instantiated is an *essential* feature of any natural property.

[24] Lewis himself (2009, 211) considers this possibility but he finds no obvious reason to block his sceptical argument.

[25] The only way a DV-ist has to preclude this possibility is by considering the actual causal/nomic role of a natural property as its individual essence.

she can regard one and the same natural property as *both* dispositional and categorical in an *epistemically perspicuous* manner. Properties can be simultaneously dispositional and categorical (as Identity Theory suggests) in the sense that each property may have two distinct counterpart relations grounded in different ontological factors. The first counterpart relation is grounded in the similarity of causal/nomic roles and is obviously related to the way of the *de re* modal representation of properties which I call DV. In this case, the property in question is deemed dispositional. The second, distinct from the first, counterpart relation is grounded in an ontological factor (or factors) completely independent of the causal/nomic roles and is related to RNDV. In this second case, the property is deemed categorical.[26]

A second general objection to property-counterpart theory is raised by Dustin Locke (2012). According to Locke, it is useless to appeal to counterpart theory as a theory of *de re* modal representation of natural properties, because, at some level of analysis, we *have to* assume the truth of transworld identity theory anyway. Hence it is wiser to use the latter right from the beginning. Locke raises this objection in order to undermine all counterpart theories of *de re* modal representation of properties. But why does he think that sooner or later we'll have to appeal to a kind of transworld identity? Locke examines Lewis' theory in order to show that invoking counterpart theory beyond the level of individuals generates a regress which must stop somewhere by assuming the transworld identity of some relevant feature. It is true that the attempt to analyse and present the grounds of counterpart relations at each level is an infinite task that threatens the very success of the process of *de re* modal representation in general. For instance, if we ask 'what is it for two counterparts to "share" a property P?', the proper counterpart-theoretic answer is 'the first must have P and the other must have a counterpart of P'. But what is it, in turn, that makes one property a counterpart of another? Suppose we posit a second-order feature of properties to ground the counterpart relation at this level. If we ask 'what is it for two property-counterparts

[26] I am not sympathetic to the widely held belief that there is an intimate relationship between DV and dispositional properties and RNDV and categorical properties, respectively (see Sect. 4.4). This does not mean, however, that those identity theorists who do hold this belief cannot utilise counterpart theory for their own ends.

to “share” this second order feature?’, the proper counterpart-theoretic response is ‘the first must have this feature and the other must have a counterpart of it’. One may go on and ask ‘what is it that makes one second-order feature a counterpart of another?’. If she posits a third-order feature to ground the relevant counterpart relation, she engages in an infinite regress. Locke thinks that if we want to stop the regress we can do it at any level we want by insisting that sharing a feature must be taken literally as a case of transworld identity. In his opinion, this move had better be taken at the second level where we have property-sharing between counterparts of individuals.

The success of Locke’s move is due to the nature of the identity relation which presumably renders redundant any attempt to find the grounds of its application. But, crucially, one can reach the same result (i.e., stop the regress) by insisting that the counterpart relation at some level is *primitive*. That is, by claiming that it is a modal *brute* fact that some higher-order features are counterparts of one another. (As we’ll soon see, this is what happens according to one of the versions of the property-counterpart theory which I claim is consistent with RNDV.) Hence, it is not the case that counterpart theory of natural properties *presupposes* transworld identity theory. It just presupposes, like all theories, the introduction of some primitives.[27]

The discussion so far does not intend, of course, to be a thorough defence of property-counterpart theory. The issue is too complicated to be settled here. Yet, I think that by addressing (adequately, I hope) the above worries, I’ve increased the plausibility of the counterpart-theoretic framework as an alternative scheme for the *de re* modal representation of natural properties. In what follows, I’ll complete my suggested solution to the permutation problem by presenting two versions of property-counterpart theory which are both (under certain assumptions) consistent with RNDV. But before that, a potential complication needs to be discussed. It is well known that counterpart relations need not be transitive. It might be objected, therefore, that RNDV-ists can avoid the permutation difficulty simply by utilising the non-transitivity

[27] The question about the legitimacy of positing the *specific* primitives is another matter that will be discussed in Sect. 4.3.4.

of counterpart relations without also endorsing the world-boundedness of natural properties. This can be done because modal claims about actual transworld natural properties need not be evaluated by considering those very properties in possible worlds in which they exist; adoption of the counterpart-theoretic framework allows the evaluation of modal claims by examining the counterparts of those properties (which, crucially, need not be the properties themselves). An RNDV-ist may exploit the fact that counterpart relations need not be transitive in order to avoid the consequences of the Permutation Argument expressed in Chisholm's form. Consider, for instance, two fundamental natural properties P and Q and a series of possible worlds from (the actual) w_1 to w_n such that: (a) in each transition from w_i to w_{i+1} ($i = 1, 2, \ldots, n-1$) a slight variation of the causal/nomic roles of P and Q occurs; and (b) in w_n, P and Q have finally swapped their actual roles. Now, while the actual P in w_1 has a counterpart in w_2, and that counterpart has in turn a counterpart in w_3, and so on, until we reach w_n, it does not follow that P has a counterpart in w_n. Given that the same is true for Q as well, it follows that, in some cases and according to counterpart theory, the swapping of the roles of P and Q is not a *genuine* metaphysical possibility. One may wonder, then, why an RNDV-ist should opt for the pair <world-boundedness of properties, counterpart theory> rather than counterpart theory alone. The answer, I think, is the following: The most reasonable motivation for choosing counterpart theory alone is that it allows a form of *essentialism*, at least in some contexts. The failure of transitivity of some counterpart relations *effectively* constrains the metaphysical possibilities of fundamental natural properties. And the most plausible reason for this choice is the assumption that, at least in some representational contexts, the causal/nomic roles (or parts of them, if that makes sense) are essential features of fundamental natural properties. I do not think that the RNDV-ist would in any case like to think of those roles as essential. So I suggest that she should both adopt (a) the world-boundedness of natural properties, in order to render the swapping scenario literally impossible, and (b) property-counterpart theory, so that, given the world-boundedness, to allow modal judgements about the possible behaviour of fundamental natural properties.

4.3.4 Two Versions of Property-Counterpart Theory Consistent with RNDV

Taking the counterpart-theoretic framework for fundamental natural properties for granted, I'll now proceed and present two RNDV-friendly versions of property-counterpart theory which can provide a solution to the permutation difficulty. In what follows, I assume that a proponent of RNDV has identified and selected (probably from the list presented in Sect. 4.2) one or more ontological factors as the ground for the *de re* modal representation of a fundamental natural property P. I refer to that ontological factor(s)—which, by definition, is completely independent from the causal/nomic role of P—simply as 'the non-qualitative factor(s) of P'.

4.3.4.1 PCT_1 (First Version of an RNDV-Friendly Property-Counterpart Theory)

As a first attempt to find a version of the property-counterpart view consistent with RNDV, let us consider an account based on the following two assumptions:

First Assumption All first-order fundamental natural properties are world-bound, while their second-order features (causal/nomic roles and non-qualitative ontological factors) are transworld entities.
Second Assumption Any actual first-order fundamental natural property P is *de re* modally represented in various possible worlds by having counterparts in those worlds; these counterparts literally share the same second-order non-qualitative factor(s) which is (are) the sole ground(s) of the *de re* modal representation of P.

PCT_1 is in a certain sense a kind of analogue of Lewis' original counterpart theory for individuals suitably adapted to fit the case of properties. In Lewis' theory, world-bound individuals are characterised by transworld properties, whereas, in PCT_1, world-bound natural first-order properties are 'characterised' by transworld second-order features. There is, however, an element of disanalogy; contrary to the spirit of Lewis' orthodox counterpart theory for individuals, in PCT_1 there is no counterpart relation based on *qualitative* features (i.e., related to causal/nomic roles) of natural properties. According to what I have already noted, PCT_1 can clearly provide an RNDV-friendly solution to the permutation problem. By accepting the world-boundedness of fundamental natural properties, it renders the swapping scenario literally impossible; and by being a kind of property-counterpart theory, it certainly allows modal judgements about the possible behaviour of fundamental natural properties. Furthermore, it is consistent with RNDV, because it acknowledges non-qualitative factors only as the sole grounds of the *de re* modal representation of natural properties. Yet, the plausibility of the suggested solution is undermined by the fact that PCT_1 faces two serious difficulties. The first one has been pointed out in another context by Lewis (1986, 229–230) and concerns the central claim of PCT_1 according to which there are no qualitative counterpart relations between properties. I think that this difficulty can be addressed and I'll discuss it in the sequel (see Sect. 4.3.4.2). The second problem, however, is more difficult to address and consequently, as far as I can see, it seriously undermines PCT_1. The origin of the difficulty is the *pluralist* stance of PCT_1 as far as the modal character of properties is concerned. First-order natural properties are assumed to be world-bound, while second-order ones are assumed to be transworld entities. The question that naturally arises is whether there is a relevant *ontological* difference which may justify the different treatment of properties and, consequently, the adoption of pluralism. The prima facie obvious answer is that the properties under consideration are of different orders. But it is not clear (at least to me) whether this fact makes a difference relevant to the modal character of natural properties. As far as I know, there are no arguments so far that can settle the issue. Perhaps one may skip the problem of ontological justification and present the pluralism of PCT_1 as a natural consequence of the alleged indispensability of transworld identity theory. Following Locke's argument I discussed above, one must accept the transworld identity (and, consequently, the transworld existence) of those features which ground the *de re* representation of first-order properties in order to avoid an infinite regress. Yet, as I have already remarked, the worrisome regress can be arrested without appealing to any kind of transworld identity. In any case, the burden is on the advocate of PCT_1 to show that the pluralistic stance is adequately justified. In absence of a relevant argumentation, I suggest that we should reject PCT_1 and look for another version of property-counterpart theory.

4.3.4.2 PCT_2 (Second Version of an RNDV-Friendly Property-Counterpart Theory)

The moral we should draw from Sect. 4.3.4.1 is that we should avoid any account that is based on a *disjointed* theory of modal representation of natural properties. The account to be presented now is based on a *unified* view concerning the *de re* modal representation of properties of different orders. Its core assumptions are the following:

First Assumption Besides the first-order fundamental natural properties, their second-order features are world-bound as well.
Second Assumption Each member of the class of world-bound counterparts which corresponds to any given fundamental natural property P exemplifies a counterpart(s) of the non-qualitative ontological factor(s) of P.
Third Assumption Any actual first-order fundamental natural property P is *de re* modally represented in a possible world by having a counterpart in this world which exemplifies a counterpart(s) of the non-qualitative ontological factor(s) of P.
Final Assumption The relation holding between the world-bound second-order non-qualitative ontological factors is modally *primitive*; that is, there is nothing that can determine which world-bound second-order ontological factors belong to the class of counterparts of the ontological factor(s) of P.

Just like PCT_1 (and for exactly the same reasons), PCT_2 can provide an RNDV-friendly solution to the permutation problem. The plausibility, however, of the suggested solution depends on whether PCT_2 can adequately address objections related to its final defining assumption. The formulation of PCT_2 assumes modally primitive counterpart relations between world-bound second-order ontological features. Doesn't this fact constitute a reason to reject it? It is certainly a commonplace that every theory has its own primitives. Therefore, if introducing a primitive is deemed a drawback for a particular theory, this must be due to the nature of the primitive itself. Perhaps, to return to our case, counterpart relations are not the sort of entities that we can easily think of as unanalysable. For instance, Lewis (1986, 229–230) insists that counterpart relations must be relations of qualitative similarity under various respects.[28] His remarks, of course, concern counterpart relations between individuals but can be easily generalised in order to apply in the case of properties as well. According to him (and using the neutral term 'entity' to refer to both individuals and properties), when two entities are non-qualitative counterparts they should stand in a certain relation or share a certain property or are both included as parts of a certain transworld mereological sum. But given that *any* two entities stand in *infinitely* many relations, share an infinite number of properties, and so on, the defender of non-qualitative counterparts must single out some of them as the proper candidates for non-qualitative counterpart relations. Lewis thinks that the haecceitist (the RNDV-ist in our case) cannot do that,

> leaving it entirely mysterious what it could mean to say that [entities] were non-qualitative counterparts. (1986, 229)

There is an *ad hominem* reaction to Lewis' claims; we can point out that Lewis too has to accept equally objectionable primitives in order to achieve his aims. For instance, when pushed on what makes certain similarities perfectly natural—a very important issue for his metaphysical edifice—he is forced to take naturalness to be primitive. As far as I can see, taking naturalness as primitive is no less mysterious than taking non-qualitative counterpart relations to be primitive. (In any case, the burden is on Lewis to show that his primitive is non-objectionable.) Regardless of the above remarks, there is a second reaction to Lewis' claims that is

[28] As I have already remarked, Lewis' objection is relevant to the plausibility of PCT_1 as well. Hence addressing it removes a serious obstacle for adopting this view. Yet, PCT_1 is objectionable due to its dependence on a disjointed account of the modal representation of natural properties.

preferable. We may grant Lewis that non-qualitative counterpart relations are difficult to grasp. Due to the exclusion of all qualitative factors, it is impossible to provide any analysis of them. But, *pace* Lewis, that fact does not compel any property theorist (and the RNDV-ist in particular) to deny the existence of non-qualitative counterpart relations. We cannot always analyse (in terms of concepts referring to other entities) our ontological posits, but this does not imply that their introduction is unjustified. Inter alia, we are justified to introduce unanalysable (primitive) entities into our ontology to the extent that their introduction is forced on us for reasons of theoretical adequacy. The case of primitively understood non-qualitative counterpart relations is an example of such an introduction. *Pace* Lewis, we must insist that counterpart relations of natural properties need not be relations of similarity of causal/nomic roles (recall once again my restrictive sense of qualitativeness in the case of natural properties). The reason is that there are modal contexts in which Lewis' qualitative approach proves to be theoretically inadequate since it cannot accommodate genuine possibilities. As a first example, though arguably a controversial one, consider a possible world which represents a natural property as having qualitative features entirely different from those it has in the actual world. By definition, qualitative counterpart relations cannot provide the means to accommodate this possibility. We can imagine, however, an example of a counterpart relation that is not qualitative and does the job. In particular, we may first stipulate[29] a *de re* possibility which represents the given actual property characterised by entirely different causal/nomic roles; and then ontologically justify our stipulation by assuming the existence of a non-qualitative counterpart relation that grounds the possibility. As I noted, this case is controversial, because the defenders of DV insist (in line with their doctrine) that causal/nomic roles are *essential* features of natural properties. We can certainly imagine, they claim, the above-mentioned possibility, but imaginability is not an infallible guide to metaphysical possibility. We cannot have a cross-world variation of essential features; hence, the above scenario does not describe a genuine metaphysical pos-

[29] The discussion concerning cases of possibility-stipulation has its origin in Kripke's (1980) work. For a brief discussion of their relevance to Lewis' counterpart theory, see Bricker (2008).

sibility and, consequently, does not justify the appeal to a non-qualitative counterpart relation. There is, however, another example which does not beg the question against DV. Both RNDV-ists and DV-ists may acknowledge the 'twin' possibility emerging from the intraworld duplication of a causal/nomic profile for any actual natural property. Granting the possibility of two distinct natural properties having the same causal/nomic role, we may easily generate non-qualitative differences between distinct possible worlds. More precisely, we can represent a possibility consistent with the above by claiming that there is a possible world in which one of the duplicate properties exists alone and a different world in which the other duplicate property exists alone. Those two worlds, by definition, do not differ qualitatively, but they do differ in what they represent *de re* concerning the two duplicate properties. It seems, therefore, that counterpart theorists (about properties) need non-qualitative counterpart relations between natural properties to accommodate the 'twin' possibility.[30]

It might be objected, however, that a property-counterpart theorist may accommodate the 'twin' possibility without admitting the existence of non-qualitative counterpart relations. She may simply follow the Lewisian strategy (1986, 230) of accommodating haecceitistic possibilities and adjust it to fit the property case. She may admit that, according to the 'twin' scenario, there are two possibilities that do indeed differ only in representing *de re* those duplicate natural properties. But she may deny that this fact implies that there are two *worlds* that differ non-qualitatively. Instead, she may claim that there are two possibilities represented by a *single* possible world. This alleged world represents one natural property twice by having twin counterparts under the same or different counterpart relations. Hence, it seems that the property-counterpart theorist can have the best of both worlds; she can respect the modal intuitions of the 'twin' scenario while not at the same time allowing a non-qualitative difference of possible worlds. But can she avoid the introduction of non-qualitative counterpart relations? To see that she cannot, consider a problem raised by Mackie (2006, 85–91) in order

[30] A DV-ist may avoid the difficulty related to the 'twin' possibility only by embracing the view that causal/nomic roles constitute the *individual essences* of fundamental properties.

to challenge Lewis' strategy in seeking to reconcile his anti-haecceitism about possible worlds with certain haecceitistic intuitions concerning the possibility of qualitatively indiscernible individuals. Mackie asks how the modal separation of haecceitistic possibilities within one possible world is achieved. She wonders why, as Lewis claims, the twin counterparts represent two *distinct* possibilities within a world. Her challenge (properly modified) can be transferred to the aforementioned case concerning the possibility of two natural properties with identical causal/nomic roles. Lewis' strategy (modified and then implemented in the case under discussion) presupposes that a natural property may have multiple counterparts in the world in which it exists. The relevant question is how, following Lewis' suggestion, the modal separation of non-qualitative possibilities within one possible world is achieved. The most plausible way to answer the question is to exploit the flexibility of Lewis' theory concerning the choice of counterpart relations. Perhaps, if we appeal to the appropriate qualitative counterpart relation(s) between the 'twin' counterparts, the latter can be modally distinguished. We may choose either the same relation for all counterparts or a different relation, one for each pair. Following the first option, however, won't do the job. Granted that, by definition, qualitative counterparts have the same causal/nomic roles, no counterpart relation based on the latter can modally distinguish them. Following the second option won't do the job either, for the same reason stated in the above sentence. If different counterpart relations are exclusively based on causal/nomic roles, they cannot distinguish 'twin' counterparts. It seems then that only by adopting non-qualitative primitive counterpart relations one can use Lewis' strategy to adequately meet the 'twin' scenario challenge. To recap: PCT_2 is based on four substantial modal assumptions. The first three are relatively uncontroversial, given that the world-boundedness of natural properties is a possible metaphysical view and the appeal to counterpart theory is legitimate even in the context of properties. The fourth assumption (the one concerning the existence of primitive non-qualitative counterpart relations) is the most controversial. I argue, however, that, far from being ontologically 'suspect', non-qualitative counterpart relations between natural properties are an indispensable ingredient of any theoretically adequate property-counterpart framework. Hence, Lewis' objections fail to undermine the assumption, and PCT_2 remains a viable account capable of offering the appropriate ground for a consistent with RNDV counterpart-theoretic response to the permutation difficulty. To the extent that, besides the permutation argument, there are no other cogent reasons against RNDV, the latter is still a live option for property theorists.

4.4 RNDV as the *Unique* Way of the *de re* Modal Representation of Fundamental Properties

If the arguments of the preceding section are sound, RNDV is a viable alternative account of the *de re* modal representation of fundamental properties. In this section, I wish to take a step further and claim that RNDV is the *unique* account. In what follows, I present the reasons for my claim.

Traditionally, there is an intimate connection between the dispositional/categorical nature of a property and its mode of *de re* modal representation. It is widely assumed that dispositional properties are *de re* represented via DV and categorical properties are *de re* represented via RNDV. I'll call this view the *Intimate Relationship* (*IR*) assumption. *Conceptually*, this exclusive relationship cannot, however, be justified because the two issues are clearly distinct. There is nothing that might compel us to think of the ontological factor 'responsible' for the dispositional/categorical nature of a property as the ground for its *de re* modal representation as well. Hence, I suggest that the core intuition about the dispositional/categorical distinction must be dissociated from the account of the *de re* modal representation of properties. Crucially for my purposes, this suggestion is supported by the symmetry-based considerations that 'flow' from modern physics and ground a specific version of RNDV. Recall that even 'paradigmatically dispositional properties', such as mass, spin and charge, can be *de re* modally represented following Option 4 which *is* a version of RNDV.[31]

[31] The existence of a scientifically based version of RNDV undermines the popular objection that the latter appeals exclusively to *controversial* ontological posits (such as quiddities and P-substrata)

As far as I can see, the aforementioned disentanglement removes the only strong reason we have to insist that there are *two* distinct types of *de re* modal representation of fundamental properties. If the exclusive relationship between DV and dispositional properties (and RNDV and categorical properties, respectively) is missing, then what prevents us from embracing the view that there is only *one* way of *de re* modal representation of fundamental properties? Indeed, application of Occam's razor in this case suggests that we *should* hold the unique-way view. But why not think (following dispositionalists) that DV is the unique way? The reason is that I think that DV faces serious difficulties. It is not only that we can easily imagine possible worlds in which actual fundamental properties confer on their bearers causal powers different from the ones they actually confer. Perhaps this intuition can be seriously undermined by embracing the view that imaginability does not entail metaphysical possibility. We have, in addition, scientifically based reasons to question the *correctness* of DV. As I argued earlier in this chapter, we may plausibly suggest that invariance-features can ground the *de re* modal representation of 'paradigmatically dispositional properties'. Even if we suppose that such invariance-features do not *exhaustively* ground the *de re* modal representation of the properties in question, there are no cogent reasons to exclude them from being *partial* grounds. Granted that, we have a direct refutation of DV according to which the *de re* modal representation of any fundamental physical property is *exhaustively* grounded in its causal/nomic roles.

Besides, we have an extra reason to reject DV. Psillos (2012) presents a scenario which clearly shows that, besides causal/nomic[32] roles, we need extra 'stuff' in order to have an adequate account of *de re* modal representation of all fundamental features. He urges us to consider two causally/nomically indiscernible possible worlds, w_1 and w_2, inhabited by *different* properties. More precisely, in w_1 there are two properties A and B acting in tandem to produce a certain causal/nomic profile Q. Crucially, A or B, taken individually, do not have (in w_1) any extra causal/nomic role

as the sole grounds of *de re* modal representation of fundamental properties.

[32] Psillos refers to the *causal* roles of properties. His arguments, however, can be extended to cover their *nomic* roles as well.

and, furthermore, do not exist in world w_2. The latter world differs from w_1 just in the fact that in place of A and B has a *single* property C characterised by Q. Psillos plausibly claims that, in contradistinction to the advocates of DV, we should recognise w_1 and w_2 as *distinct metaphysically possible* worlds. And this can be done only if we assume that the grounds of the *de re* modal representation of properties are not exhausted by their causal/nomic roles.

Finally, it might be objected that my previous remarks do not compel one to admit that RNDV is the unique way of *de re* modal representation of properties. There is always the possibility of following a *moderate* RNDV—an account according to which *both* causal/nomic roles and non-qualitative ontological features are grounds of *de re* representation. But once again, in this case, Occam's razor dictates that we should rest content just with non-qualitative characteristics.

5

Further Objections to Categorical Monism

In the previous chapter I argued that categorical monists have the resources to meet the allegedly fatal objection from the unpalatable consequences of Quidditism (RNDV, in my terminology). It is now time to discuss some further objections to Categorical Monism.

5.1 The Argument from the Truthmakers of Unmanifested Dispositions

A relatively popular line of argument against Categorical Monism is to show that categorical properties (by themselves or in tandem with laws of nature) are not sufficient truthmakers for the truths related to *unmanifested* dispositions of elementary fundamental objects. Note that the problem is not that categorical monists cannot offer *any* truthmakers for that special case of modal truths. Rather, the problem is supposed to be that categorical monists cannot offer *sufficient* truthmakers for unmanifested dispositions. Since many philosophers regard this argument as the most convincing one, in what follows, I'll examine its merits thoroughly. As we'll see, however, it is not clear whether we do need inherently modal

V. Livanios, *Science in Metaphysics*,
DOI 10.1007/978-3-319-41291-7_5

properties (i.e., dispositional ones) as truthmakers for the truths concerning unmanifested dispositions.

In what follows, I'll concentrate on Armstrong's categorical monistic account. For Lewis, another prominent categorical monist, the truthmaker for an unmanifested disposition of an object is a state of affairs which obtains in *concrete* possible worlds 'close' to the actual one. This state of affairs consists in a counterpart of the object displaying the manifestation under consideration (in response to an appropriate stimulus). I find Lewis' appeal to concrete possibilia far-fetched and so I'll ignore it in the following discussion. Let us begin, then, with the case of unmanifested dispositions in *nomically possible* contexts, that is, in the actual world and in all possible worlds with the same nomic web as ours. According to one interpretation of the notion of manifestation of a disposition (see, for the specific proposal, Esfeld and Sachse (2011)), fundamental objects of our world (and plausibly of other worlds physically possible with respect to the actual one) *continuously* manifest their dispositions by creating and maintaining relevant fields of forces. Prima facie, according to this account, categorical monists have nothing to worry about because eventually there are *no* unmanifested dispositions related to fundamental features either of the actual world or of nomically possible worlds.[1] (Esfeld and Sachse think that all fundamental features are dispositional; nevertheless, categorical monists may use their suggestion in order to avoid the problem that allegedly besets their view while denying that those properties are dispositional.) In spite of its initial appeal, however, this account can be directly challenged. Within the classical field-theoretic context and under a realistic conception of fields, the phrase 'The field here-now is such-and-such' is made true by an ascription of a possibly unmanifested dispositional property either to a sui generis entity (the field itself) or to the spatiotemporal location which the field 'occupies'. The possible manifestation of the dispositional property is expressed by

[1] There exist at least two other notions of manifestation according to which there *cannot* be any unmanifested dispositions. According to the first notion, dispositional properties are always conjunctive properties one part of which is the property instantiated at the manifestation event. The second view construes dispositional properties as four-dimensional structural properties which 'contain' the manifestation-property as a temporal part. One might object to both interpretations precisely because of their incompatibility with the 'commonplace' that there are unmanifested dispositional properties. For details, see Tugby (2013).

the counterfactual 'Had a body of a specific kind been here-now, it would have felt a specific kind of force of such-and-such value'.[2,3] The case in the quantum-field-theoretic context is far more complex due to the various formulations of QFT and the difficulty to clarify the ontological commitments of each distinct formulation. Yet, at least for free fields and despite the disagreements about how to interpret the Operator-Valued-Quantum-Field (OVQF),[4] the latter can be characterised

> in terms of complex dispositions associated with the space-time points—and most importantly, *dispositional* relations among the points. (Teller 2002, 160, my emphasis)

Most of these dispositions are unmanifested in the sense that only through combining OVQF with a contingently occurring physical state can we get an arrangement of manifest expectation values for relevant quantities at various space–time points and the correlations among them. It seems possible then to claim that the truthmakers for these unmanifested dispositions are genuine dispositional properties and relations. Of course, as I have already remarked, the issue is far more complex and I do not pretend here that I have even scratched the surface of the difficulties involved. What I have tried to show, however, is that the appeal to the notion of a continuously manifested field does not obviously help the categorical monist to cope with the truthmaking problem about unmanifested dispositions of fundamental properties.

A second, more orthodox, view about manifestation associates it not with the fields related to fundamental properties but with fundamental

---

[2] According to a non-realistic conception, talking about fields at specific locations is a convenient manner to describe possible effects of the spatial arrangement of distant property-possessors at earlier times. For a discussion of the various interpretations of the classical notion of field, see Lange (2002, Ch.2).

[3] Here is another challenge to the continuous-manifestation suggestion: Suppose that gravitational mass is a fundamental property of elementary objects. One might then claim that an elementary particle having that property is always manifesting gravitational attraction. But as Chakravartty (2013) rightly points out, this fact holds only at the *determinable* level. The *determinate* magnitude of the attraction depends on changeable relations to other particles and one cannot argue that *all* determinate dispositions of gravitational attraction are continuously manifested.

[4] See Paul Teller (1995) and the contributions of Andrew Wayne, Gordon Fleming and Teller himself in the Kuhlmann et al. (2002) volume dedicated to the ontological aspects of QFT.

interactions. In this context, one may appeal to the following scenario (presented by Tooley (1977) who modified the initial suggestion of Martin (1997, 203)[5]) to challenge Categorical Monism. Consider a world populated by ten types of fundamental particle. Given that each particle may interact both with particles of the same and different types, the world in question may be governed by 55 interaction laws. Suppose now that all but one of those laws are known; the interaction law of x-type with y-type of particles is, however, not known. This is due to the fact, that, though this type of interaction is physically possible (i.e., it conforms to the web of actual laws), the boundary conditions of our universe are such that omnitemporally no x-particle is appropriately related to a y-particle in order to interact with it.[6] Tooley argues that in this case we have good reasons to assert the existence of an uninstantiated law governing x–y interactions and the associated unmanifested dispositions of x- and y-particles to interact in a *specific* way. Armstrong (1983, 119) proposes that we should offer an account of such 'uninstantiated laws' and unmanifested dispositions by appealing to counterfactuals about what laws *would* hold if, contrary to fact, the proper stimulus condition were activated (i.e., x- and y-particles had a proper relation R; the occurring of the condition is tantamount to the instantiation of the associated structural universal xRy). Tooley-Martin's challenge to the categorical monist is to offer sufficient truthmakers for the counterfactual truths concerning the possible interaction between x- and y-particles. Generally, there are two available options: The first is to claim that the required truthmakers are just the fundamental properties of elementary objects which are (in order to be able to play their truthmaking role *by themselves*) irreducibly dispositional, inherently modal, properties. Armstrong, *qua* categorical monist, wishes to avoid this option since it stands in direct contrast to his view. The second option is to insist that properties *plus* laws are truth-

[5] Martin notes that he had already devised the scenario in the mid-50s and used it against the reductive accounts of causal dispositions.

[6] The assumption that grounds Tooley's possibility, namely that some particles (albeit not whole kinds of them) omnitemporally do not interact with each other, is consistent with recent scientific findings in cosmology. For instance, the causal structure of GR allows for cosmological models with *particle horizons* in which different portions of the universe cannot even exchange signals with one another (see Wald (1984, §5.3); for a philosophical discussion, Earman (1995, Ch.5)).

makers sufficient for the counterfactual truths in question. Here, one may choose either dispositional or categorical properties as constituents of those truthmakers. The first choice is not available to the categorical monist, so she turns unavoidably to the second one. Yet even if she opts for the second alternative, there is room for two different views. The first is Tooley's Factual Platonism, according to which the proper truthmakers for the aforementioned truths contain *uninstantiated* properties and laws. Armstrong (rightly in my view) rejects Tooley's solution because it is against the spirit of Naturalism. The only view left is one according to which only *instantiated* properties and laws can be regarded as sufficient truthmakers for the unmanifested dispositions of the Tooley-Martin case. Armstrong suggests that the appropriate strategy here is to follow a course akin to the one he proposes in order to solve the problem of 'missing' values in functional laws. According to Armstrong's (1983, 112) solution for the 'missing'-values-problem, statements of laws corresponding to the 'missing' values of functional laws should be construed as *counterfactuals* about what laws would hold if, contrary to fact, certain universals (corresponding to the 'missing' values) were instantiated. The truthmakers for such counterfactual truths are instantiated higher-order functional laws which govern a set of lower-order laws (*some* of which are instantiated) plus the instantiated universals (corresponding to the 'non-missing' values). Implementing this strategy in Tooley-Martin's case, we have to assume the existence of a higher-order law governing the x–y interaction which, however, does not yield *definite* results as in the analogous case of functional laws. Tooley thinks that this disanalogy creates problems for Armstrong's view. Armstrong argues for the contrary using the following example of an irreducibly probabilistic, disjunctive law. Consider the irreducibly disjunctive law: If P, then Q or R (not both), where Q and R are equiprobable. Consider also the counterfactual: If a at t had had property P, then Q or R would have resulted. The counterfactual holds but there would seem to be no truth of the matter about which of Q and R would have occurred. Analogously, it seems that in Tooley-Martin's scenario it is a true counterfactual that, if 'alien' and actual particles were to meet, they would have a unique mode of interaction. But we are not entitled to conclude that there is some unknown, ontologically determinate mode of interaction that would have occurred. Just like the disjunctive law case (where it is ontologically indeterminate whether Q or R would have occurred), the consequent of Martin's counterfactual is ontologically indeterminate (Armstrong 1996, 92). Martin responds that there is no analogy between Armstrong's law and his scenario. For, Armstrong needs instantiations of the disjunctive law, which is, occurrences of Q and R, while Martin does not need repeated occurrences as instantiations of *any* law whatsoever (*ibid*., 138). Armstrong counters that Martin's scenario requires that, apart from the unknown interaction, *other* types of interaction should occur, and in each of those other cases a law peculiar to that case is involved. Armstrong relies on this fact to justify his choice for the truthmaker Martin asks for (*ibid*., 151). Here it seems that we eventually have reached an impasse, and nothing so far has shown that Armstrong's categorical monistic suggestion is not able to provide a sufficient truthmaker for Tooley-Martin's unmanifested dispositions.

The case of unmanifested dispositions of objects in *nomically impossible* contexts is a more controversial issue. On the one hand, nomically impossible contexts can be created when there are *actual* laws dictating that the antecedent of the counterfactual associated with the unmanifested disposition cannot occur in our world (or in worlds with the same web of laws as ours). On the other hand, nomically impossible contexts may also be created in possible worlds with *alien* laws. This latter case is related to the *metaphysical* possibility of unmanifested dispositions. It might be objected that, though Armstrong's account can offer sufficient truthmakers for *actual* and *nomically possible* unmanifested dispositions, it is unable to do so for *metaphysically possible* ones. (Even according to the manifestation-as-creation-of-fields view, there is nothing that might exclude the *metaphysical* possibility of fundamental properties not creating fields. In other words, even continuously manifested properties are not *necessarily* manifested. The metaphysical possibility of field-less unmanifested dispositional properties is furthermore consistent with the basic intuition—which for many is definitional for dispositional properties—that dispositional properties can exist unmanifested.) This is true, but it is at least doubtful that it has any force against Categorical Monism as the correct metaphysical account of the nature of fundamental properties in the *actual* world. For, as Armstrong points out, even if such dispositions are possible, they are *merely* possible. According to him, it is not reasonable, on the basis of such non-actual, metaphysically possible cases, to postulate the existence of irreducible dispositional properties since the actual and the nomically possible facts are (arguably) well accommodated by the truthmakers suggested by categorical monists.

Though I think that Armstrong's reaction to the metaphysical possibility of unmanifested dispositions is well justified, here I would like to tentatively suggest a sketch of a response to the relevant difficulty in the spirit of his solution to the 'missing'-values problem for functional laws. To this end, let me first present Armstrong's most recent (1997, 82) account of what a truthmaker for an unmanifested disposition of an object would look like in the categorical monistic context. Consider an object o having the power or disposition to M (instantiate the manifestation property M) when condition S (the stimulus property S) is present. For Armstrong, the manifestation M (when it occurs) is just the instantiation of the consequent of the law C&S → M, where C is a categorical property possessed by o. So, the truthmaker for the manifested disposition is the state of affairs <o having C> plus the law <(C&S) R_N M>, where R_N is the (contingent) nomic necessitation relation. In the case where the manifestation M does not occur (even necessarily), Armstrong proposes that the truthmaker is still the state of affairs <o having C> plus *all* laws involving property C. Laws are needed for two reasons: first, because the state of affairs <o having C> does not by itself *entail* that the object o has the disposition to M when S. And second, because the false (in this case) assumption that the state of affairs <o having S> obtains must be empirically possible, that is, it must be compatible with the actual laws of nature.[7] Consider now the object o and the case where o has the disposition M due to an *alien* law. (Recall that categoricalists most often insist that laws are metaphysically contingent and, consequently, that there are possible worlds populated by the actual properties but with laws different from the actual ones.[8]) A statement about such an unmanifested disposi-

[7] Armstrong's suggestion can be applied to Lewis' (categorical monistic) framework as well; the laws, however, should be defined in this case according to the best-system account. Nevertheless, Lewis prefers to use his concrete possibilia in order to provide truthmakers for the actually unmanifested dispositions.

[8] In his (2004a, 136), Armstrong deviates from orthodoxy and suggests that the relation of nomic necessitation between categorical properties must be metaphysically necessary. His new view is

tion due to an alien law could be construed as a *counterlegal* about what manifestation would occur if that alien law (instead of the relevant actual one) were a law of the actual world. And the truthmaker for that counterlegal could be an *interworld* 'higher-order' law which 'governs' a set of 'lower-order' *world-bound* laws (including the actual and the alien ones) plus the actually obtaining state of affairs <o having C>. The proposed truthmaker is in a sense an entity of the actual world since one 'part' of its interworld constituent (i.e., the relevant actual law) exists at the actual world. Following this suggestion, a categorical monist may (arguably) provide an adequate *this-wordly* truthmaker for the truth expressed by the counterlegal under consideration.

Returning now to the case of nomically possible contexts, one might wonder what the problem with the case of unmanifested dispositions really is. We might think that Armstrong's aforementioned account fails because categorical monists cannot plausibly claim that two *different* truths have the *same* truthmaker. But in fact they do not have the same truthmaker; in the manifested case the truthmaker consists in the state of affairs <o having C> plus only one law. Given the actual obtaining of <o having C>, we do not need the other laws involving C to guarantee its empirical possibility. Perhaps the problem is that categorical monists offer the same *kind* of truthmaker both for non-modal and modal truths. But it cannot be; if we bypass Lewis' appeal to possibilia inhabiting concrete possible worlds, the truthmakers for modal truths of our world must be found among the entities of the actual world. Armstrong does exactly that; he uses actual categorical properties plus actual laws for truthmakers. Dispositionalists also do that by appealing to inherently modal fundamental properties. So why is it a *special* problem for categorical monists?

Eventually, are Armstrong's truthmakers sufficient or not? Do we need in the case of unmanifested dispositions property-truthmakers which are intrinsically modal? According to one approach, the answer to those questions depends on our opinion regarding the *existential status* of the dispositions in question. Armstrong himself points out that his proposal yields a deflationary 'soft' ontological account of unmanifested disposi-

based on the endorsement of (a modified and extended version of) Donald Baxter's (2001) theory of instantiation as partial identity between universals and particulars.

tions. He insists that his account is not an *eliminativist* one, but it surely is an account that ontologically *downgrades* the unmanifested dispositions of objects. For the latter *supervene* on the suggested truthmakers and, according to his notion of supervenience, they should be no addition of being. Armstrong thinks that he has an overwhelming reason for avoiding any ontologically *robust* notion of unmanifested dispositions. This is the well-known Meinongian Objection; unmanifested dispositions seem to be *somehow* related to non-actually existing events or states of affairs, their manifestations. Hence, by reifying dispositional possibilities we are committed to the existence of Meinongian objects which is an anathema for any naturalistic philosophy.

It is not clear whether Armstrong's objection poses an intractable difficulty for the dispositional realist since there exist various responses to it. Some dispositionalists (Heil (2003) and Martin (2008)) try to 'dissolve' the problem by insisting that there is no *genuine relation* holding between a dispositional property and its non-occurring manifestation.[9] Others, instead, choose to think of dispositional directedness as a *genuine* relation and offer alternative solutions to the problem. Philosophers like Ellis (2001, 132–5), Mumford (2004, 192–5) and Tugby (2013) hold the view that the directedness relation relates universals[10] and claim that their account has the resources to overcome the Meinongian objection. To show why, consider first what a proponent of *transcendent* universals can reply to Armstrong. According to her view, universals can actually exist *uninstantiated* and so the manifestation-universal actually exists *even if it has no actual instances*. The directedness relation, therefore, has always two actually existing relata and the possible lack of actual instances of manifestation poses no threat to Dispositional Realism. An advocate of *immanent* universals can also avoid the Meinongian objection, at least for some dispositional properties. The basic principle of immanent uni-

[9] This view has, however, the obvious drawback that it leaves *ontologically brute* the dispositional-directedness-to-a-mere-possible-event fact.

[10] This is not to say that there is no escape route for the trope theorist as well. She may appeal to *actual abstract unrealised* possibilia in order to ensure that the directedness relation has two actually existing relata even if the dispositional property is unmanifested (see Bird (2006) for the relevant proposal in a universalist context). Here, I find Tugby's (2013) criticism persuasive; positing unrealised possibilia makes genuine dispositional properties explanatorily *redundant* as truthmakers for various modal/causal truths.

versalism is that a universal cannot exist 'outside' spacetime and it always 'lives' *in its entirety* in each of its instances. Hence the only thing that a dispositional realist needs to ensure that the directedness relation has two actually existing relata is that the manifestation-universal has at least *one* instance in a spacetime region which may be arbitrarily far away from the particular instance of the dispositional property.[11]

It seems then that there is no insurmountable obstacle to adopting ontologically *robust* unmanifested dispositions. Yet, do we *have to* include them in the actual ontological inventory? Here, I am sympathetic towards Armstrong's view which, in a nutshell, tells us that the whole issue has to do with a *practical* attitude of human beings and has nothing to do with ontology proper. In his words (1997, 82):

> They [human beings] have to represent the various possibilities to themselves, to image them perhaps, and in general to invest mental energy in them. If they do not do all this, it seems, they lack the emotional and intellectual *drive* to think about these possibilities effectively. It helps in this planning to *reify* dispositional possibilities. (My emphasis)

Nevertheless, a number of dispositional realists think that the 'reification' of unmanifested powers (more precisely, their upgraded existential status) of elementary fundamental objects is not just a corollary of a practical attitude. It is rather strongly *suggested* by science itself. To this end, they have developed 'arguments from science' to defend the genuine existence of fundamental dispositional properties.

5.2 The Argument from Science: Scientific Practice

Recently, a battery of arguments for the existence of genuine dispositional properties has been provided based on scientific methodology and practice. These arguments have taken various forms such as *transcenden-*

[11] The immanent universalist's response has to deal with the Tooley-Martin scenario. How can the immanent universalist cope with this case, where by definition there can be *no* manifestation of the dispositional properties under consideration?

tal reasoning, inference to the *best* explanation or even inference to the *unique* explanation. For instance, in the former form, the relevant arguments are based on an uncontroversial claim about the practices scientists engage in and a transcendental premise asserting that in order for the first claim to be possible/comprehensible/intelligible Dispositional Realism must hold. The conclusion is that Dispositional Realism is actually true. Differently formulated, the aforementioned arguments are intended to prove that Dispositional Realism provides the best or even the unique explanation of the success of some aspects of scientific methodology.

A lot of ink has been spilled on whether such transcendental arguments or inferences to the best explanation can achieve the ontological aim they are designed for. Consider, for instance, Cartwright's attempt to defend the existence of dispositional properties at the fundamental level on explanatory grounds. Cartwright (2009, 155) claims that in some cases scientists have to appeal to unmanifested dispositions in order to provide the best explanation of specific scientific facts. A favourite example for her claim is the case of two elementary particles (of the same kind) that remain unmoved though they attract and repel each other due to the possession of different fundamental properties (e.g., mass and electric charge). Cartwright thinks that it is absurd to hold that in that case we have actually manifested properties. But whether it is indeed bizarre or not depends on our preferred notion of manifestation. Cartwright obviously assumes that the disposition associated with each property is manifested through the *motion* (better, acceleration) of particles. But this is not the only available conception of manifestation. Alternatively, we may claim that both properties of both particles are manifested to the extent that the relevant fields have been created and maintained. Or we may take the total *null* effect on each particle as a kind of 'maskish' manifestation of one of its properties due to the presence of the other. The possibility of non-unmanifested-disposition interpretations undermines Cartwright's claim that we *need* to assume the existence of genuine dispositional properties in order to explain what actually happens in the two-particle scenario. Even if dispositional properties are the only kind of properties that can be triggered without manifesting themselves, it is not sure that they are present in Cartwright's case. Based on the above-mentioned alternative notions of manifestation we may offer metaphysically explanatory accounts of Cartwright's scientific fact without positing unmanifested dispositional properties. And, as far as I can see, it is not clear why her explanation is the *best* one.

After this example, let us now briefly examine two aspects of scientific practice the success of which allegedly provides grounds for believing in genuine dispositional properties. The first one is the *analytic method* as applied to physical science. Implementing this method, physicists can explain the behaviour of a *complex* physical system by first identifying its relevant parts (sub-systems), then conducting experiments (most often under simple, 'shielded' conditions) to learn their behaviour, and finally using this knowledge to explain the behaviour of the complex system. Corry (2009) discusses the issue and tries to draw ontological conclusions from the successful application of the method. He claims that in order for the method to be successful (and even intelligible) we must presume the existence of an ontological component associated with the parts of the complex system which remains *invariant* when these parts are put together. After examining the prospects of several Humean candidates (and finding all of them wanting), he considers the case of Cartwright's 'open-ended' capacities as a case of a serious contender for filling the required 'invariant' role. According to Cartwright (1999), capacities related to fundamental dispositional properties remain invariant in different circumstances though they may give rise to a great variety of behaviours. Corry stresses the fact that, if dispositional realists adopt a two-layer view of dispositional properties and the observable behaviours they give rise to as their manifestations, they cannot possibly explain why scientists have any warrant to apply the analytic method. For, there is no invariance of the behaviour in different circumstances in which properties may find themselves. Cartwright posits an intermediate ontological level to avoid the difficulty. In her terminology, this level contains *exercises* of natural capacities. That level is in fact (and *pace* Cartwright who rejects their existence) populated by the (component) *forces* associated with each fundamental property. Armed with this extra ontology, a dispositional realist may find in the intelligibility and the success of the analytic method in physics an argument for her view. In particular, she may claim that dispositional properties are invariant[12] dispositions to 'produce' forces which, in their turn, remain invariant when we shift from simple conditions to more complex ones related to compound physical systems. Of course, for the analytic method to succeed, the dispositional realist needs also a law of *composition* of forces (to get, in each case, the resultant force) and an extra law (such as Newton's second law) to associate a manifest behaviour (e.g., acceleration) with the resultant force. One might have qualms about the consistency of the above-posited laws with the basic tenet of orthodox dispositionalism according to which all fundamental laws 'flow' from dispositional essences of the fundamental properties. The crucial (for the present purposes) point, however, is that we do *not* have to assume the existence of irreducible *dispositional* properties in order for the analytic method to work. In fact, given the existence of component forces as the invariant elements required for the success of the method, one may insist that it is *categorical* properties in tandem with laws that give rise to them. As Corry (2009, 180) points out, we should only posit an ontological level of genuine dispositional properties if we think (for independent reasons, as Corry himself does) that the dispositions for producing forces are irreducibly dispositional. Hence, the argument from the success of the analytic method does not add any extra arrow to the dispositional realist's quiver.

Another aspect of scientific practice that arguably leads to the adoption of an ontology of genuine dispositional properties is, in Cartwright's terminology, the method of *generalisability*. Scientists conduct experiments in highly specialised conditions (isolated physical systems), express the results in the form of laws and then generalise and apply the latter to different circumstances (where the physical systems are not isolated). According to a number of arguments, the commitment to the existence of genuine dispositional properties (and perhaps to robust causal laws, either universal in scope, or 'local') makes possible/intelligible the above practice and explains in the best (or even in unique) manner its success. Philosophers like Bhaskar (1975), Cartwright (1999) and Hüttemann (1998; 2007), despite their differences, all appeal to such arguments to

[12] In any situation where the dispositions in question are triggered, they will give rise to the relevant forces.

justify their preference to dispositional ontology. Just like the already examined case of the analytic method, the success (and the intelligibility) of the method of generalisability seems to demand an ontological element which can be invariant when scientists shift from ideal to non-ideal conditions. Given that genuine dispositional properties give rise to different manifest behaviours in different circumstances, we need (just like in the case of the analytic method) the intermediate level of component forces to secure the required invariant ontological factor. And once again the problem a dispositional realist must address is that positing a level of irreducible *dispositional* properties which produce these forces seems to be unwarranted.[13]

We may also consider the case at hand from the perspective of laws of nature. Think, to begin with, that laws aim to describe/predict the (regular) *behaviour* of certain physical systems. Given that, we must admit that that behaviour cannot be the *manifest* one. For, as Hüttemann (1998; 2007) points out, laws *describe/predict* the behaviour of physical systems in mostly *unrealised* circumstances. (To relate to the above remarks, that could be because they often concern the component forces, not the resultant ones.) Laws, however, can be *applied* to actually, more complex, realised situations as well. How *can* that be? Hüttemann thinks that the most plausible answer to this question is that laws ascribe to physical systems *dispositions* for specific manifest behaviours. But, as Chakravartty (forthcoming) also remarks, application is not guaranteed; we cannot

[13] We have to clearly separate the issue of whether one is entitled to assume the existence of genuine dispositional *properties* from the related case concerning the actual existence of *any* dispositional element. It is at least debatable, however, whether one can ascribe a dispositional or a categorical character to any entities except properties. Be that as it may, a dispositional realist may claim that the argument from scientific practice can at least prove the indispensable existence of *dispositional forces*. I disagree, but I am not going to dwell on that here. I'll just point out a difference in the two arguments presented in this section. In my view, the case of scientific generalisability seems to create a *less* favourable context for the dispositional realist. To illustrate, note that, in the case of the analytic method, a dispositional realist may claim that her argument from scientific practice, though unable to prove the existence of genuine dispositional properties, can nevertheless show that an irreducible dispositional element must be introduced in our ontology. Component forces can be prima facie regarded as *capacities* for manifest behaviours of actual physical systems which are parts of other compound systems. In the case of scientific generalisability, however, one cannot claim that; for, even if it is assumed that component forces are capacities, they can only be associated with manifest behaviours of systems in highly specific 'shielded' conditions not occurring outside laboratories (the case is even worse in scientific abstraction, where we deal with *ideal* physical systems).

without argument simply claim that the conditions in actual complex circumstances are such that they guarantee the applicability of laws describing dispositions manifestable in different circumstances. For that reason, Hüttemann introduces the notion of Continuously Manifestable Disposition (CMD): CMDs are dispositions that can be *partially* manifest, while the transition from the 'ideal' situation to the conditions of *complete* manifestation (in the actual complex circumstances) is smooth. The behaviour of a physical system possessing a CMD is a continuous function of the degree to which the manifestation conditions are actually realised. The application of a law is ensured because it [the application] does not require the systems possessing the nomically relevant CMDs to display their manifestations *completely*. Of course, as Hüttemann (2007) acknowledges, if we want to *explain* the continuity between ideal and non-ideal circumstances for the manifestation of a CMD, we must have both a complete list of the possible interference factors and a relevant law of composition which describes (or determines) how those factors affect the 'ideal' behaviour of a physical system possessing the CMD. Regardless of whether Hüttemann's suggestion can provide a reasonable ontological ground for the success (and intelligibility) of scientific generalisability, it does not prove that CMDs must be construed as *irreducibly* dispositional and so 'flow' from genuine dispositional properties that actual physical systems possess.

Consider now the possibility that laws do not even in principle aim to describe/predict the regular behaviour of physical systems. Smith (2002) and Earman et al. (2002) suggest that there is a crucial distinction between laws and differential equations describing the temporal evolution and motion of physical systems. The motivation behind this proposal is, as Smith (*ibid.*, 244) explains, based on the

> ability to model *different* situations using the *same* set of laws according to a modeling recipe which allows for the derivation of concrete differential equations of motion.

According to this view, laws such as Coulomb's law do not (aim to) tell us anything about the manifest behaviour of charged bodies. If we want information about the latter, we must appeal to the differential equations derived by combining nomic facts (Coulomb's law + Newton's second law) and non-nomic assumptions (for instance, that the bodies in question are not massive). It is the involvement of the latter that explains why the differential equations of motion are *not* laws; they depend on non-nomic assumptions which are necessary in order to apply a certain physical theory to a specific actual case. Laws concern the component forces 'produced' by fundamental properties. Ontological commitment to these forces does not entail, however, that there is a level of irreducible dispositional properties which 'generate' them. It may well be true that laws of forces 'enforce' objects possessing *categorical* properties to 'produce' specific component forces.

5.3 The Argument from Science: The Scientific Characterisation of Fundamental Properties

For some dispositional realists, one of the main arguments supporting their view is that the very nature of *scientific characterisation* of the fundamental features of our world indicates that the latter are dispositional. More precisely, they claim that the *whole* set of fundamental properties that physical science attributes to elementary particles seem to be scientifically characterised *only* in terms of what (causal) powers bestow on their bearers and *therefore* are dispositional. In Mumford (2006) we have a clear exposition of this belief:

> [The premise that] the properties of subatomic particles are (all) dispositional is supported by physical theory, not just as it is interpreted by philosophers but also by scientists disinterested in this (i.e. dispositional vs categorical) debate. [This premise] attributes properties to subatomic particles – spin, charge and mass – that *all* appear dispositional. (2006, 475)

We can find similar remarks in the works of other dispositionalists such as Ellis (2002, 47) and Molnar (2003, 135). Though popular, I find this kind of argument inconclusive for three reasons. First, because there are actual fundamental relations (the spatiotemporal ones) which confer no causal powers on their bearers and so cannot be characterised in the above (causal) manner (for a defence, see Sect. 1.3.1.4). Second, because some of the fundamental properties (such as mass, spin and electric charge which, as we saw, are presented by dispositional realists as clearly dispositional) *can* be characterised independently of the causal powers they bestow on their bearers (for details, see Sect 4.2). And finally, because even if *all* fundamental properties and relations were causally conducive and exclusively characterised by their causal roles, this would not entail that they are all dispositional. To prove this, one should argue that each of these properties is a constituent of an appropriate state of affairs which is *by itself* (part of) a minimal truthmaker for the truths associated with its causal roles. Yet, as far as I can see, the appeal to the manner of scientific characterisation (even if it is ubiquitous) cannot directly prove the truth of this (definitional for the dispositional properties) claim.

Surprisingly, even Armstrong (a prominent categorical monist) claims (in his (1997, 76–9)) that the argument from the 'dispositional' scientific characterisation of fundamental properties raises difficulties for Categorical Monism. In opposition to dispositional realists, however, he restricts the argument to the case of properties mathematically characterised as *vectors*. After defending the view that vectorial quantities are genuine properties attributed to objects at certain instants of time he claims that we may plausibly construe those properties as dispositional because

> we seem to have no handle on these properties except via their [causal] effects. (1997, 78)

Armstrong's remarks are brief at this point, but I think that his argument rests on the following three premises:

(1) Vectorial properties are *identified* in terms of what they do, in terms of their [causal] effects.
(2) It *seems* then that vectorial properties are exclusively *de re* modally represented by DV.
(3) The IR assumption (namely, that dispositional properties are *de re* represented via DV and categorical properties are *de re* represented via RNDV).

Given all that, he arrives at the conclusion:

∴ Vectorial properties can be plausibly construed as dispositional properties.

Pace Armstrong, I think that the argument from vectorial quantities raises no serious difficulty for the categorical monist. First, it is unclear whether the identification mentioned in (1) is intraworld or interworld. If the former, there is no reason to assume that vectorial properties are dispositional. For categorical properties can be also intrawordly identified in terms of what they (albeit in tandem with the nomic relations) 'do'. Second, we have scientifically based reasons to challenge the claim that *all* vectorial (more generally, non-scalar) properties are exclusively *de re* modally represented by DV. For instance, it can be shown that, in the context of classical field theory, both the four-momentum and the angular momentum of a field can be identified as conserved quantities due to the invariance of the relevant action under space–time translations and spatial rotations, respectively. Repeating my remarks from Sect. 4.2, and given that the intraworld identity of these fundamental physical properties can be supplied in the actual world via symmetry considerations, nothing prevents one from suggesting that the associated second-order invariance-features can ground the *de re* modal representation of those properties too. Of course, this does not prove that *all* non-scalar quantities can be thus identified. Nevertheless, it does cast doubt on the assumption that there is a *special* intimate relationship between dispositionality and the non-scalar 'character' of some fundamental properties. Finally, as far as I can see, we may also plausibly challenge the third premise of the argument; the IR assumption about the correspondence between the dispositional/categorical nature of a property and its mode of *de re* modal representation cannot be conceptually justified since the two issues are clearly distinct. As I have already remarked, there is nothing to compel us to think of the ontological factor 'responsible' for the dispositional/categorical nature of a property as the ground for its *de re* modal representation as well. For all these reasons, I think that Armstrong's argument from vectorial quantities is far from convincing.

5.4 In No Categorical Terms?

In a recent work, Kerry McKenzie (2014) considers the prospects of a categorical monistic view about fundamental properties of the actual world. She (just like the majority of metaphysicians) relates the answer to that question to the issue concerning the modal status of laws of nature. In the sequel, I'll articulate (and find wanting) an argument against categorical properties based on her remarks.

(1) Categorical properties (by definition) can be relata in nomic relations which satisfy fundamental symmetries entirely different from the actual ones.
(2) The behaviour of the duplicates of the actual objects (of a natural kind which is supposed to be defined by a set of categorical properties those objects instantiate) in other possible worlds can be described/governed by laws which satisfy entirely different fundamental symmetries from those of the actual one.
(3) In each possible world inhabited by duplicates of the actual kinds of objects the fundamental symmetries of the actual world hold.
(4) The fact asserted in (3) constitutes a non-trivial structural constraint on candidate laws which may govern/describe the behaviour of objects belonging to the actual natural kinds.
∴ The natural kinds to which actual objects belong cannot be defined by any set of categorical properties.

The first premise is not controversial since, by definition, categorical properties can be related by different nomic relations in other possible worlds. Among those alien nomic relations may be some that do not satisfy/obey actual fundamental symmetries. The second premise is not controversial either since categorical properties by their very nature do not necessarily bestow a specific behaviour on their bearers. Hence, in other possible worlds, the behaviour of objects which share with the actual ones all of their qualitative properties can be described by laws entirely different from the actual ones. Again, among those alien laws may be some that do not satisfy/obey actual fundamental symmetries.

Let me now turn to the third premise. It is grounded in the following syllogism: The fundamental laws of the actual world concern the fundamental interactions in which elementary particles participate. According to the Standard Model of fundamental interactions, fundamental laws are associated with specific symmetries which classify the elementary particles obeying those laws into families with members possessing the same value of mass, but different values of the same determinable properties. The mathematical theory of Lie algebras (which describes the action and properties of fundamental symmetry groups) suggests that each family of elementary particles corresponds *necessarily* to one and only one symmetry. Hence, as (3) asserts, in each possible world inhabited by duplicates of the actual kinds of objects, the fundamental symmetries of the actual world hold.

Unlike the first three premises, (4) seems controversial. McKenzie does not prima facie offer cogent reasons for the truth of (4) though she assumes that we confine the whole discussion to possible worlds inhabited by elementary objects belonging to the *actual* natural kinds (and, hence, by (3), the actual fundamental symmetries hold). As she herself acknowledges, it is not true in general that a law of nature can be *uniquely* determined by the symmetries associated with it. McKenzie thinks, however, that unique determination is actually achieved in the case of *fundamental* laws. Her claim is based on the following assumptions:

(a) Theories aiming to provide a description of the fundamental features of the world must be valid up to *arbitrary high* energy (infinite energy limit).

The constraint expressed in (a) follows inter alia from the spacetime *continuity* required by Lorentz invariance. Given, however, a number of findings in work related to some candidates for the proper quantum theory of gravity, it is not clear whether the continuity of spacetime is a fundamental feature of the actual world.

(b) The only *quantum field-theoretic* laws that exist in the infinite energy limit are those which are asymptotically free.[14]

Here the basic assumption that makes this fact relevant to McKenzie's argumentation is that in each possible world inhabited by elementary particles belonging to the actual kinds, the *quantum field-theoretic* laws hold. But do we have to *assume* that? McKenzie's answer to this question is '*tu quoque*'; for, according to her, within the context of the 'canonical'[15] debate on the modal status of laws of nature a *specific* form of laws is *in fact* assumed, that is, the *functional* form.[16] The functional form of laws, however, is not a crucial element of the arguments expressed within the 'canonical' debate (most probably, its use is a matter of convenience and perhaps simplicity). What *really* matters is the metaphysical hypothesis that laws express (in some way or another) relationships among natural properties/relations which are not necessarily functional in form. In that sense, the whole debate does not even presuppose a *determinate* answer to the question concerning which forms of laws are possible (and how metaphysicians may know something about it). Hence, the debate enjoys an air of independence of the various scientific approaches about laws, but simultaneously makes it extremely difficult (impossible?) to determine the specific nomic variations which may be regarded as metaphysically possible.

Regardless of the adequacy of the above remarks, the objector may continue: Do we have to assume that *quantum field-theoretic* laws hold in each possible world inhabited by elementary particles belonging to the actual kinds? Since McKenzie understands the objection in a context of *contrast* between classical and quantum-mechanical laws, her response to

[14] Asymptotically free theories are the ones the coupling constants of which have zero limit as energy scale tends to infinity. Characteristic example is Quantum Chromodynamics, the theory describing the strong interaction between elementary particles. Due to its asymptotic freedom, quarks behave almost like *free* particles when interacting at high energy levels.

[15] In McKenzie's sense, the 'canonical' debate (with some notable exceptions) is taking place within the context of a 'canonical account' which associates contingentism about laws with categoricalism about properties and necessitarianism about laws with Dispositional Essentialism.

[16] For McKenzie, a functional law is a law of the form $a(x) = f(b_i)$, where b_i are real- (or real vector-) valued functions representing determinable physical properties and relations, and f is some functional (i.e., a function of functions).

this objection is that it is not clear whether there exists a possible world governed by the laws of classical physics at the fundamental level. This can be true but is also irrelevant for the present discussion; for the objection does not point to the possibility of fundamental *classical* laws, but to the possibility of *non-quantum-mechanical* laws instead. So the issue really concerns the question whether there is a possible world governed at the fundamental level by such laws. As we'll see below, McKenzie's further assumptions rule out such a possibility.

(c) The only asymptotically free quantum theories in *four dimensions* are the so-called renormalisable local gauge theories.
(d) On the assumption that the fields concerned are specified, the laws of such theories are uniquely specified *but for the values of the constants* appearing in them.

This final assumption clearly shows that *strictly speaking* there is no unique determination of laws by symmetries even assuming the truth of all the other assumptions. The truth of (4), however, does not require the *unique* determination of laws by symmetries. It is sufficient for symmetries to impose non-trivial interworld constraints on laws. And *that* is true, *given however that the laws in question concern the behaviour of objects which belong to interworldly* ***fixed natural kinds*** (in our case, to the actual natural kinds). Yet, the laws of a possible world relate *properties* of *that* world; if we want to check whether the *same* properties (or their counterparts) can be related by different laws in different worlds, we should not have to assume that they form *interworldly fixed clusters.* In other words, the interrelated debates concerning the dispositional/categorical distinction and the modal status of laws respectively are framed in terms of *properties* (and perhaps their counterparts in other possible worlds), *not* in terms of *duplicates of* (elementary) *objects.* Indeed, had the debates been framed in terms of duplicates of objects, a genuine metaphysical possibility would have been hidden. In my view, there is a (metaphysically) possible world populated by the actual (or counterparts of) state-independent fundamental properties (clusters of which define the actual natural kinds) in which *no* actual natural kinds exist. This world is metaphysically possible because it is characterised by actual symmetries (such as the Poincaré symmetry) which provide the identity of state-independent properties, but not by the actual *internal* symmetries which provide identity conditions for particle kinds. McKenzie claims that before we start to discuss the issue of the modal status of laws we should first agree on the *kinds* of objects that fundamental laws may govern. Yet, the demand to delineate the context of discussion by fixing the kinds of objects to the actual (quantum-mechanically described) ones (as she insists) *confines* the possibility space to those worlds characterised by the *whole set* of actual symmetries. We only get constraints that allegedly prove the non-categoricality of fundamental properties if we put *by hand* other constraints related to the overall symmetry-structure which a world may instantiate.

It seems, therefore, that we cannot reach conclusions about the *de re* modality of properties and laws from the premises of the argument we discussed above. More precisely, the argument does not entail that the Humean metaphysical package of categorical properties and metaphysically contingent laws is untenable within the context of modern physics. Given what McKenzie says, I conclude that she has her own qualms about the efficiency of any kind of argument like the above against the Humean framework, but for another, entirely different, reason. In particular, she suggests that the nomic necessity allegedly proved by the argument is purely *mathematical*; it is a mathematical fact concerning the relation between Lie algebras and the associated multiplets of definite values of specific state-independent fundamental properties. Under that perspective, the alleged nomic necessity is eventually acceptable to Humeans who have no problem to admit a necessary 'relation of ideas'. In that case, the appeal to the findings of contemporary physics modifies the conditions under which the discussion on a pure metaphysical issue is taking place, without however purporting to be a crucial factor for the evaluation of the truth value of any metaphysical view concerning the world-structure.

I am sympathetic to the view that scientific findings may be sometimes unrelated to the *truth* of a metaphysical view concerning the structure of reality. But if we want to be science-informed metaphysicians, we have to admit that the appeal to science can at least *enlighten* the associated metaphysical problem. In McKenzie's case, however, instead of having an illumination of the *de re* problems of categoricality and nomic contingency by physics, what we really have (according to her suggestion) is the resolute decision that in the case at hand the mathematical part of our best physical theories can only inform us about our concepts and not about the world. In fact, McKenzie's suggestion fits nicely a Strawsonian *descriptive* metaphysical view, because it construes the upshot of the argument as indicating a necessary connection between concepts comprising our *conceptual* structure to describe the world. (Perhaps, this could be the answer for most of my objections; our conceptual scheme compels us to assume some fundamental truths in all possible worlds.) Yet, for most contemporary philosophers, metaphysics in general deals with what the world is (or would be) like. Hence, I think, that the most reasonable stance the vast majority of science-informed metaphysicians should adopt regarding the best mature physical theories is a broadly realistic one; specifically those philosophers should take seriously the possibility that physical theories reflect (or, at least, have something to say about) the world-structure itself.

To avoid misunderstandings, I am not claiming here that McKenzie's approach is *incompatible* with science-informed metaphysics. Even advocates of the latter could accept that *some* scientific findings might shed no light to metaphysical issues regarding the structure of the world simply because they concern the conceptual means we use to describe the world. The problem with the case at hand, however, is that mathematics is so intimately involved in physical theories that a science-informed metaphysician may find it extremely difficult to see the latter realistically, while claiming simultaneously that the mathematically expressed explanatory relations appearing in them do not reflect (or, at least, have something to say about) the structure of the world *at all.*

6

The Contingent Character of Categoricality and Dispositionality

Thus far I have defended a kind of Categorical Monism for the fundamental features of the actual world. The overall argumentation rests mainly on a posteriori evidence related to recent scientific findings and practice. In this chapter I discuss what I think is an important corollary of following the methodology of science-sensitive (or science-based, if you prefer) metaphysics used in this work. This consequence has to do with the *modal strength* of the metaphysical views about the nature of actual fundamental features. More precisely, in my view, *following such a methodology commits the metaphysician to the view that whatever metaphysical theory about the nature of actual fundamental features is (actually) true it is only* ***contingently*** *so.* Let us see why.

6.1 The Metaphysical Contingency of Categorical Monism

Consider the following question:

(DETERMINATION) *How can the precise determination of the elements that constitute the truthmakers for the modal truths related to fundamental properties and laws be achieved?*

V. Livanios, *Science in Metaphysics*,
DOI 10.1007/978-3-319-41291-7_6

The first response to DETERMINATION is to claim that physics is *not* the unique provider of the elements that constitute the required truthmakers. Defenders of this view emphasise the need for providing truthmakers for modal truths concerning *alien* properties and laws and point out that, *exactly* because it discovers *only* the properties and laws of the actual world, physics cannot *by itself* provide all the required truthmakers. In that case, however, what is the *extra* source from which the required truthmakers 'flow'? The most popular answer is modal *intuitions*. (Though there is no consensus among philosophers for their proper account, I think we may agree, for the purposes of this work, that modal intuitions can be construed as unreflective and non-inferential judgments about modal issues.)

The second response to DETERMINATION, which is par excellence compatible with a science-informed metaphysics, is to claim that the set of all mature, most successful theories of contemporary physics is the *unique* provider of the required elements by offering (ideally) the full inventory of fundamental properties, laws and their actual interrelationships; those elements *fully* determine the truthmakers of *all* modal truths under consideration. One way to defend such a view is to insist that there are *no* possible worlds populated by *alien* properties and laws. The latter belief can be considered either as a consequence of the *nature* of actual fundamental features (see, for instance, Bird's strong necessitarianism about laws) or as a *brute* modal fact. An alternative way is to admit that there exist possible worlds inhabited by alien properties and laws, but, at the same time, claim that the only legitimate and non-futile (as far as the progress of metaphysics is concerned) way to discover truthmakers for modal truths concerning entities of those 'alien' worlds is to appeal to the fundamental features and laws of *our* world.[1]

Now it seems that an important corollary of following the second response is that the distinction between nomological (aka physical) modality (in the sense of consistency with the actual nomic web) and metaphysical modality is eliminated (or, at best, blurred). In this sense,

[1] For French and Mckenzie (2012), anyone who insists on investigating the behaviour of physical entities in such 'alien' worlds totally deprived of all available means (i.e., for them, the findings of fundamental physical theories), she should either *postulate* that some specific possibilities obtain or adopt an *agnostic* stance towards the issue.

there is no room left for any kind of *autonomous* metaphysical modality concerning the actual fundamental features and laws. On the contrary, due to its appeal to modal intuitions, the first response has the resources to establish a significant degree of autonomy for metaphysical modality. Construed as an a priori method, the appeal to intuitions has been regarded as the *exclusive* field of a metaphysical research, which thus determines the *nomologically impossible* worlds and consequently delineates the context of a metaphysical possibility that lies between nomological and logical possibility.

In what follows, I'll first argue that a science-sensitive metaphysician *should* acknowledge a kind of autonomous metaphysical *possibility*. Let me begin by examining the first response. A preliminary objection is that the science-sensitive metaphysics *rules out* right from the start the appeal to intuitions related to common sense. Yet, that is not true; consider, for instance, the interrelated metaphysical issues of the persistence of material objects and change. It seems that the Theory of Relativity lends support to those theories of persistence (like *perdurantism* and *exdurantism*) according to which all material objects have temporal parts.[2] However, for a number of science-informed metaphysicians those theories have a great metaphysical *cost*: they arguably abandon our *intuition* of change as alteration. As four-dimensional objects, all material things cannot gain or lose any property. Change consists in the generation and destruction of temporal parts which instantiate incompatible properties.[3] This conflict with a significant metaphysical intuition motivates in part a search for a proper compromise between it and the physics-informed theories of persistence. Another recent example concerns the case of the gauge-invariant interpretation of GR and its importance for the formulation of adequate quantum gravity theories. The aforementioned interpretation yields a *frozen time* formalism which stands in conflict with our basic intuition (based on our experience) that there is *real* change in the world. In particular, endorsing the assumption that the only physically real quantities in GR are those which are gauge-invariant, we are soon led to the

[2] For a brief presentation of the relevant arguments and bibliography, see Hawley (2010).

[3] For a comprehensive collection of articles examining all the relevant issues concerning persistence and change, see Haslanger and Kurtz (2006).

conclusion that, for any model of the theory, there is no real quantity which takes on different values in distinct times; in other words, there is no real change in the world provided we follow such a (at least allowable) construal of the best theory we currently have about gravitation and spacetime. Once again, the opposition between the intuitive acceptance of time flow/associated change and the physics-based *Parmenidean* metaphysical possibility has motivated various approaches ranging from the adoption of the frozen time view (and the concomitant attempt to explain why everyday experience and experimental observations seem to involve genuine change) to the full endorsement of a Heraclitean view of permanent change.[4]

Nevertheless, the appeal to common-sense-based intuitions has been criticised for a variety of reasons. The majority of critics have challenged the *reliability* of the intuitions-based methodology of doing metaphysics. In particular, it has been argued that modal intuitions exclusively related to common sense are (due to their historical and evolutionary determination) unreliable for judging issues concerning entities far remote from common experience, while their totality most probably forms an *inconsistent* set of beliefs. Another common complaint is that intuitions tend to differ among philosophers and that there are no clear criteria establishing who the expert on a given issue is. Furthermore, as Humphreys (2013) points out, the reliability of intuitions is also *domain-specific* and consequently varies even for the same philosopher. An unsuccessful answer to the above difficulties is to appeal to intuitions based *only* in scientific facts. This move actually threatens to turn the first response into the second.[5] I think (but I cannot argue here, due to limitations of space) that the whole battery of arguments against intuitions undermines the first response (to the extent that we talk about common-sense-based intuitions). It is better then to seek for a view compatible with the spirit

[4] Belot and Earman (2001) is a useful presentation of the frozen time problem and the various approaches to it.

[5] Another problematic aspect of the appeal to intuitions related to scientific facts concerns the project of providing explicit or implicit definitions of scientific concepts. Conceptual analysis is significantly complicated due to the presence of various approximations and idealisations in science. For a discussion, see Humphreys (2013).

of science-informed metaphysics *and* largely immune from the above-mentioned objections. To this end, I turn now to the second response to DETERMINATION.

A serious worry is that, in the context of this response, we (*qua* science-sensitive metaphysicians) have *no* motivation to search for the truthmakers related to alien entities. And given that, we cannot motivate in turn the introduction of a kind of metaphysical possibility distinct from the nomological one. I do not agree; consider the modal truths associated with subjunctive conditionals (a) having antecedents opposed to actual laws (i.e., counterlegals); (b) having antecedents which contain laws with different from the actual values of fundamental constants (which current physics cannot fully determine); and (c) that assume certain combinations of laws and symmetries which, though not holding in the actual world, can nevertheless be regarded as possible in the sense that the mature, most successful theories of contemporary physics *allow* them. In all the above cases (which are perfectly respectable under the scientific point of view), we need truthmakers related to truths concerning *alien* nomic webs. I do not deny that we should appeal to the fundamental features and the elements of the nomic web of *our* world to do that. I just point out that we should acknowledge the need for possible worlds *beyond* those governed by the *actual nomic web* (i.e., the totality of all actual laws, *symmetries* and their [actual] *interrelationships*). Crucially, these nomologically *impossible* worlds can be defined without invoking common-sense-based modal intuitions, but only with the help of the following (two versions of the) notion of *weak* compatibility between the metaphysical possibilities and the findings and/or practice of the best available science:

(a) A possibility is weakly$_1$ compatible with scientific findings about (and/or scientific practice based on truths concerning) the actual nomic web if it represents possible worlds in which only *some* of the nomic features of the actual world hold.
(b) A possibility is weakly$_2$ compatible with scientific findings about (and/or scientific practice based on truths concerning) the actual

nomic web if it represents possible worlds in which specific *relationships* among elements of the actual nomic web do not exist.[6]

All worlds weakly compatible with physics in the above sense are obviously nomologically impossible, but, in my view, possible in *metaphysical* sense. Yet, why should any science-sensitive metaphysician embrace the view that these worlds are possible in *any* sense? Under a (neo)Humean perspective, the answer is quick: Humeans insist that actual fundamental features are categorical and consequently may be governed by laws which differ from the ones of our world. Hence, they implicitly endorse the metaphysical contingency of the actual nomic web. But what about those science-informed anti-Humean metaphysicians who think that laws 'flow' from the dispositional essences of fundamental properties? As I shall argue in Sect. 7.2, *conservation* laws cannot be plausibly accommodated by this view, so its follower has no cogent reason to think that worlds inhabited by the actual fundamental features, *but without a specific conservation law*, are metaphysically impossible.[7] Furthermore, she may concede (since contemporary fundamental physical theories *allow* it) the existence of worlds in which specific relationships among elements of the actual nomic web do not exist. For instance, in the actual world, a specific relationship of co-existence holds between global and local gauge symmetries. To my knowledge, however, the structure of the Standard Model (which is currently the most successful account of fundamental interactions) does not *dictate* that such a relationship holds *necessarily*. Thus, the possibility remains of a world characterised by a global symmetry only, not accompanied by the corresponding local one.[8] In general, I conclude that the above-mentioned case of weak compatibility supports a conception of

[6] The two forms of weak compatibility are not independent of each other. Given the reasonable requirement that for a relation to exist in a world all of its relata should also exist in that world (and that no nomic feature is actually 'isolated'), the lack of some actual nomic features in a possible world implies that some relationships among elements of the actual nomic web do not exist in that world.

[7] For the same reason, as I argue in Sect. 7.3, she has no compelling reason to exclude worlds governed by laws with different from the actual ones values of fundamental constants (which current physics cannot fully determine).

[8] For an example of such a possibility emerging from the disconnection between global and local U(1) symmetries, see Livanios (2012b).

autonomous metaphysical possibility independent of problematic modal intuitions.[9]

A corollary of the existence of nomologically impossible but metaphysically possible worlds is this: Since scientific findings about (and scientific practice based on) truths related to fundamental properties and nomic features are obviously grounded only in the elements of the actual nomic web, they cannot be true in all *metaphysically* possible worlds; in other words, they are *metaphysically contingent*. Consider now any argument in support of a specific metaphysical view about the nature of actual fundamental features with at least some of its premises based on scientific findings and/or practice. Since the latter are metaphysically contingent, there are metaphysically possible worlds in which they do not hold. In these worlds, the premises based on them are false and, consequently, the argument is not sound. Hence, by appealing to such an argument, one cannot prove the metaphysical *necessity* of the view at issue. Since I take it that a minimal commitment of the follower of science-sensitive metaphysical methodology is inter alia to invoke this kind of arguments to support her views, I conclude that she should rest content with metaphysically contingent accounts of the nature of the actual fundamental features.[10]

Given the previous remarks, the arguments presented in this book cannot (and are not supposed to) prove that Categorical Monism is *necessarily* true. Though actually true, Categorical Monism *could* be false. Before we examine an important consequence of the metaphysical contingency of Categorical Monism, it would be useful to stress the following points. The proposition 'All actual fundamental properties are categorical' is not metaphysically contingent either because we find it possible to *imagine* its falsity (the link between imaginability and metaphysical necessity

[9] A possible worry is that the history of physics clearly indicates that sometimes nomological possibility (based on models of physical theories) is *broader* than the metaphysical one. Surely, physical theories have often expanded the range of possibility beyond both the *intuitively* possible and what philosophers have thought of as a priori possible. But this supports the above objection only provided that the metaphysical possibility at issue is conceived of as grounded in the common-sense-based modal intuitions.

[10] Of course, one might 'elevate' the modal status of her preferred view *by fiat* assuming a necessity-premise. This assumption, however, would be at odds with my upshot about the metaphysical contingency of scientific findings and practice.

is debatable) or because is known a posteriori (epistemic contingency and a posteriori knowledge are consistent with metaphysical necessity). Furthermore, I do not claim that I can know the modal status of the above proposition on a priori grounds (just as we know, one might claim, that mathematical propositions, if true, are necessarily true). What I do claim is that we know that the proposition in question is metaphysical contingent because we, *qua* followers of a science-sensitive metaphysical methodology, find ourselves incapable of proving its metaphysical necessity via arguments based on a posteriori metaphysically contingent evidence.

Embracing the contingency of Categorical Monism compels one to challenge an implicit assumption made by all property theorists sharing with me the conviction that no progress can be made in the dispositional/categorical debate unless we engage in a serious ontological discussion. In particular, all metaphysicians who hold that there is an ontological distinction between kinds of fundamental natural properties implicitly assume that properties are dispositional or categorical *necessarily*. But this cannot be true if Categorical Monism is contingently true. From the *possible* falsity of Categorical Monism follows that there is at least one possible world in which at least one actual categorical fundamental property (or its counterpart) is dispositional. Furthermore, the actual world can be considered as a possible world in which dispositional properties of other worlds are categorical. In the next section, I discuss this important, for the dialectic of the dispositional/categorical debate, issue.

6.2 Challenging Orthodoxy

The implicit assumption of the participants in the dispositional/categorical debate is that a fundamental natural property is dispositional or categorical in all possible worlds in which it exists. Furthermore, in order to give a verdict on the dispositional/categorical character of any fundamental property, metaphysicians also assume the *rigid* application of any adequate criterion of dispositionality/categoricality that can be used in the actual world.[11] Hence, they examine whether the property in question has in all possible worlds (in which it exists) the appropriate ontological features in order to be dispositional/categorical according to their preferred criterion.[12]

Of course, it comes as no surprise that the assumption under scrutiny is standard in the relevant debate. In general, it is routinely assumed that ontological distinctions (including, besides the dispositional/categorical distinction, the universal/particular distinction and the abstract/concrete distinction) refer to features which entities possess necessarily. For instance, particulars 'ought' to be particulars in each world in which they exist and there is no world in which a concrete entity exists and is abstract. Nevertheless, there are a few dissenting voices suggesting that certain entities may possess at least some of the aforementioned features contingently. Linsky and Zalta (1994), in their attempt to defend an actualist interpretation of the simplest quantified modal logic, suggest the existence of contingently non-concrete entities, while Peter Vallentyne (1998) defends the metaphysical contingency of the naturalness of properties. Finally, Fraser MacBride (1999), in the context of his examination of the prospects of modal reductionism, argues that if there are no necessary *de re* connections in nature, then we may have to countenance the possibility that any particular entity (such as Armstrong himself!) could have been a universal. With these remarks in mind, let me now return to the issue of dispositionality/categoricality of fundamental natural properties. As I have remarked, the usual procedure for (at least the vast majority of) metaphysicians—the one I do not intend to challenge—is to apply their preferred criterion of dispositionality/categoricality in all possible worlds. Notice, however, that the implementation of that criterion in every possible world does not *in general* guarantee a unique upshot. It may well be that a fundamental natural property satisfies the criterion of dispositionality/categoricality in the actual world and fails to satisfy it in another possible world. In other words, it might possess the ontological

[11] I assume that the criterion in question is ontological.

[12] For instance, according to the majority of dispositionalists, the ontological mark of dispositionality of any fundamental natural property is the necessity of its causal/nomic roles.

marks of dispositionality/categoricality *contingently*. Hence, following an unorthodox route, I suggest that one can admit the existence of fundamental *contingently* dispositional/categorical properties, provided they fail to satisfy an adequate criterion of dispositionality/categoricality in at least one possible world (whether it is the actual world or not).[13]

Two preliminary clarifications: First, one should clearly distinguish the aforementioned view from the Double-Aspect Theory once suggested by C.B. Martin (1993). In Martin's view, each natural property has (in each world in which it exists) two distinct but inseparable ontological aspects (or 'sides'), its dispositionality and categoricality, which are necessarily co-existent. According to my view, however, there are fundamental natural properties which, in each possible world in which they exist, *and in their entirety*, are either dispositional or categorical. Second, I must emphasise that I am not suggesting the common Humean view according to which the contingent laws of nature impose on any fundamental natural property a *particular* dispositional character. The thesis of Contingent Dispositionality/Categoricality (which, henceforth, I will abbreviate as ConD/C) is not a mere reformulation of the orthodox Humean view. To illustrate that, consider, for instance, the property of being electrically charged. I am not arguing that 'attracts other oppositely charged bodies' is a contingent second-order feature of the first-order property in question. Rather, my claim is that *dispositionality* (*categoricality*) itself contingently characterises this fundamental natural property.

Though contingent features of properties, dispositionality and categoricality confer *by necessity* on their bearers the nature dictated by the relevant criterion. Hence, though a dispositional property P may lack dispositionality, the latter confers a dispositional nature (as the relevant criterion determines) on P in *each* world in which P possesses it. In this respect, dispositionality and categoricality resemble the Armstrongian relational property which a property F possesses when it is nomically related to another property G. When the nomic relation N holds between F and G, F possesses the metaphysically *contingent* relational property

[13] The rigid application of the criterion in question does not entail that all possible worlds are populated only by properties of the actual world. There may exist alien properties which in *their* worlds are dispositional (categorical) according to the (relevant) criterion.

'-$_N$G' which, nevertheless, bestows *necessarily* (namely, in all worlds where F possesses it) a specific behaviour on the bearers of F. In my view, the metaphysically necessary connection that exists at the second level between dispositionality/categoricality and the natures conferred on their first-order property-bearers does not betray the spirit of the Humean rejection of necessary relations between wholly distinct existences. Just as in the Armstrongian theory of contingently necessitating nomic relations, Hume's dictum holds because there is no metaphysically necessary connection between wholly distinct *first-order* fundamental properties. Of course, due to the presence of dispositional properties in a possible world, certain *links* between distinct existences are 'created'. These links, however, are not metaphysically necessary; they have instead a restrictive intraworld modal power 'necessarily' relating instances of properties only within those worlds in which they hold (for an attempt to clarify the nature of those links in terms of forces, see Schrenk (2009; 2010), who nevertheless believes that their existence introduces a non-Humean element into the world).

6.3 Objections

I am sure that a number of philosophers may find the very idea concerning the contingency of dispositionality/categoricality rather peculiar. There are various reasons that might justify this belief. Consider first the case of natural dispositional properties, such as solubility, which are *conceptually* related to a specific causal role. ConD/C, if applicable to such a case, dictates that the property of solubility, which is dispositional in the actual world, may be categorical in another possible world. In that world, which may have radically different laws from those of the actual one, there is no assurance that solubility confers on its bearers the same causal powers which it actually confers. Since, by hypothesis, solubility is categorical in that possible world, it is the alien laws of *that* world that determine the behaviour of objects instantiating solubility. Hence, in the world in question, an actual soluble object may fail to dissolve (in the appropriate circumstances) despite the fact that it instantiates solubility. But is it not absurd to claim that objects may instantiate solubility and not dissolve in the appropriate circumstances? To meet this prima facie strong objection we must notice that dispositional properties which are by definition related to specific causal roles are not *fundamental* natural properties.[14] Consider, for instance, electric charge which is one of those fundamental properties that belong to the reduction (or supervenience) base of solubility. Charge actually confers on its bearers the power 'attracts other oppositely charged bodies', but it is not conceptually related to this specific causal behaviour. What holds for charge also holds for all fundamental natural properties acknowledged by physical science. So, to the extent that ConD/C concerns fundamental properties, the above objection raises no difficulties.

Of course, one may insist and raise an objection for the fundamental properties themselves: How can charged bodies fail to attract other oppositely charged bodies in other possible worlds? The answer is that they can, provided that we hold that the *de re* modal representation of fundamental natural properties is independent of their causal roles. If there is a genuine difficulty here, it concerns the issue of the *de re* modal representation of properties and not the intelligibility of ConD/C.[15] It is important to make clear that the issue of the acceptance of ConD/C is orthogonal to the issue of the transworld stability of a property's roles. One may hold that any fundamental natural property *must* be dispositional in *all* possible worlds (in which it, or its counterparts, exist(s)) and, nonetheless, believe that it may have different roles in different possible worlds. For instance, Hendry and Rowbottom (2009) defend the thesis of *Dispositional Contextualism*, according to which having a dispositional property is tantamount to having a single set of actual *and* possible dispositions, rather than just a set of actual dispositions. Dispositional Contextualism is a position that respects the orthodox view according

[14] They either are supervenient on or can be reduced to a web of fundamental natural properties.

[15] Returning to the case of solubility, it is reasonable to assume that the 'weird' behaviour of charged bodies in the world in question may prevent any actually soluble object to dissolve in the appropriate circumstances, and, since solubility is conceptually related to this specific behaviour, the aforementioned object may be no longer soluble. Since we cannot exclude this possibility, it seems that solubility cannot exist in a world and fail to be dispositional in it. (Recall that assuming a categorical character for solubility leaves open the possibility that objects instantiating it do not dissolve; an absurdity that gave rise to the objection in the first place.) But, as I have already remarked, this is not a problem for ConD/C; it is not assumed in the first place that, according to ConD/C, solubility (*qua* non-fundamental property) is a contingently dispositional property.

to which dispositional properties possess dispositionality necessarily; it simultaneously allows, however, a kind of transworld variation in a property's dispositional profile. Vice versa, a fundamental natural property may have the same causal roles in all possible worlds in which it exists and, nonetheless, be dispositional in some worlds and categorical in others. In the former worlds, those roles are grounded in the dispositional nature of the property itself; in the latter, they are imposed on it by the contingent laws of nature.

The claim concerning the unintelligibility of ConD/C can be also expressed in a different manner. Consider the view according to which properties are *ways* objects are; by analogy, dispositionality/categoricality must be a way a property is. The objection is that, though we can conceptually discern an object from a way that object is (or could be), the supposed distinction between ways a property is and the property itself is obscure. Saying that there are contingent ways a property is, is tantamount to saying that the property itself might have been different. And considering the case of the property that could have been different we only contemplate—would be objectors insist—a different property.[16] So, once again, how can one and the same property be dispositional in one possible world and categorical in another?

In a sense, the above objection simply begs the question against ConD/C. For, from the perspective of ConD/C, there are contingent ways a fundamental natural property could be though remaining the same. Nonetheless, as I have remarked earlier, the whole objection may rest on the issue of the *de re* modal representation of contingently characterised properties. It seems that ConD/C is indispensably committed to RNDV and, recall, dispositionalists have repeatedly argued that RNDV faces several difficulties which, of course, also beset any proposal that is committed to it. The alleged problem has also another aspect: if RNDV is presupposed, ConD/C seems to be almost trivial. If any property *can* have any causal/nomic character whatsoever, then *of course* an actually dispositional property may turn out to be categorical and vice versa.

[16] Heil (2003, 121) raises this objection for the specific case of a second-order feature F* (of a dispositional property F) which disposes F-bearers to produce bearers of another (characteristic for the manifestation of F) property G.

In order to meet this objection, one may either deny that ConD/C is indispensably committed to RNDV or argue that the latter can successfully address the objections raised against it. I've already followed the second horn and show (given some independently motivated metaphysical assumptions) that RNDV is not a problematic account after all. Here I just point out that ConD/C is not trivial even if the truth of RNDV is granted. To illustrate, suppose that you are a categorical monist and accept RNDV as an adequate position about the *de re* modal representation of fundamental natural properties. Nevertheless, in line with all categorical monists, you do not think that properties possess their categorical character contingently. You implicitly assume the necessity of the categorical nature and so you accept that natural properties can have different causal roles in different worlds but cannot have a dispositional nature in any of these worlds. This shows that in the case of categorical properties RNDV does not imply ConD/C. Now, do we have a cogent reason to assume the contrary, as far as the dispositional properties and their dispositional character are concerned? I think not and so I conclude that ConD/C is not a trivial consequence of RNDV.

Another possible objection to ConD/C is that it offers no explanation of *how* an ontological feature, such as the dispositional character, can be contingent. But what kind of *explanation* is the objector asking for? I suppose that a description of a (causal?) mechanism which explains how a property can acquire or lose its dispositionality/categoricality in different possible worlds would be enough. In this case, however, I can show that the objector's demand is exaggerated. Consider, for instance, the analogous case of the contingent features of concrete particulars. Humeans and non-Humeans alike accept the fact that there are properties which characterise concrete particulars only in some possible worlds in which they exist; most often the existence of possible worlds in which those particulars do not instantiate the aforementioned properties is grounded only in intuitions regarding what is possible or not.[17] Humeans, for instance, may

[17] To avoid misunderstandings, note at this point that I do *not* claim that, *given an initial intraworld distribution of properties*, Humeans and non-Humeans are unable to provide any means *independent of their intuitions* to specify the property distribution at a later time. Certainly, philosophers of both camps have a causal/nomic story to tell for *that* case. Rather, I am talking about the reasons they have to posit specific initial property distributions in each world they deem possible.

try to ground these intuitions by invoking a principle of recombination, the application of which 'generates' all possibilities. (Some of these possibilities are compatible with the fact of non-instantiation of contingent properties.) In this way, Humeans are able to tell a story about how the actual world (in which the instantiation takes place) differs from possible worlds in which the instantiation of the contingent features does not take place. They merely insist that the difference is due to the different global distributions of properties among the same concrete particulars (or among their counterparts). But, in any case, they do not posit a mechanism which (causally) explains the difference by generating the different distributions. I think that something analogous can be said in the case of contingent dispositionality/categoricality. It is too much to ask Humeans or anyone else to describe a mechanism which generates the relevant differences. As in the preceding case, philosophers may rely on intuitions in order to ground the relevant possibilities. Humeans may even try to reply to the objector's explanatory demand by telling a story about how worlds in which a property is, for instance, dispositional differ from worlds in which the same property is not dispositional. They may appeal to a recombination principle concerning the ontological features of properties themselves and argue that the relevant difference is due to different global distributions of these features among the same properties (or among their counterparts).

Some philosophers may object that ConD/C, though conceivable, is not viable. They might claim that for ConD/C to be viable, the relevant criterion should not *prejudge* the result of its application in all possible worlds. But how can the criterion in question not prejudge the upshot of its application in other possible worlds, given that, for instance, most of the suggested criteria of dispositionality are modal in character (hence, by definition, they involve in their application other possible worlds)? To this I can reply that the fact that all the suggested criteria of dispositionality are modal in character does not imply that the range of possible worlds which is required for the application of each criterion must be either exhaustive or the same in every possible world in which the criterion is to be applied. The criterion must be the same in all worlds, but the family of worlds used for its implementation may differ; hence, insofar as the criterion does not refer to *essential* ontological features (in which case the range of relevant worlds could be universal), a property which satisfies it in the actual world may fail to satisfy it in a possible world that does not belong to the family of worlds relevant for the application of the criterion. Hence, the result of the application of the criterion is not in general predetermined.

As I pointed out, the only case in which the criterion of dispositionality/categoricality *should* prejudge the result of its application is when it refers to *essential* ontological features. This is indeed the case with the dispositionalist's preferred criterion of dispositionality. Dispositionalists crucially entangle the features grounding the criterion of dispositionality of any fundamental dispositional property with the ontological factors grounding its transworld identity (and, in general, its *de re* modal representation). In particular, they think that the mark of dispositionality of any fundamental dispositional property is the necessity of its causal/nomic roles, while the latter are *also* essential features of the property *exclusively* constituting its identity in every possible world in which it exists.[18] Accordingly, they hold that if a property satisfies their preferred criterion in the actual world, it must satisfy it in every possible world (in which it exists), and is therefore necessarily dispositional. In other words, from the perspective of dispositionalists, the anticipated result of the application of the criterion of dispositionality is *unique* and fixed in advance.

Finally, it might be objected that ConD/C does not sufficiently address the issue of the categorical/dispositional distinction, since the latter is not a higher-order one concerning the (contingent) possession of a second-order feature, but rather the first-order one of whether or not there is something feature-like which is irreducibly modal in the actual world. In response to this objection I should first point out that I do not disagree that fundamental dispositional features are irreducibly modal, while categorical ones are not. However, I reject the conclusion that this fact implies that the categorical/dispositional distinction *should* be construed as a first-order issue. To understand the reason for my rejection, we should distinguish the sense in which first-order dispositional

[18] In terms of modal representation, causal/nomic roles are the sole grounds of the *de re* representation of any fundamental dispositional property in other possible worlds.

properties are irreducibly modal from the sense of modality I refer to when I talk about the contingent possession of dispositionality. The former is related to the physical/nomological possibilities and necessities (which, according to traditional Dispositional Realism, 'flow' from the essences of dispositional properties), while the latter are related to the kind of *metaphysical* possibility I defend in Sect. 6.1. To my mind, the first-order modal character of dispositional properties is related to the possession of the second-order feature of dispositionality. This, however, leaves room for a modality of the second kind, which gives perfect sense to the claim that properties may possess the feature of dispositionality contingently. To avoid misunderstandings, I would like to conclude this section by pointing out that my 'second-order' view does *not* imply that the physical/nomological modality is *always* related to the feature of dispositionality. Physical modality exists in worlds with no genuinely dispositional fundamental properties and is grounded in the nomic web of those worlds. Furthermore, even in worlds inhabited by fundamental dispositional properties where the physical modality is in fact related to the feature of dispositionality, my view does not imply that the former is *determined* by the latter. Only if we further assume that (a) laws and their interrelationships 'flow' from the dispositional essences of the fundamental properties, and (b) that the feature of dispositionality determines the *specific* dispositional essence of *each* (fundamental dispositional) property, we may plausibly defend a claim of determination. Yet, we have strong reasons to reject at least the first assumption (for details, see Chap. 7).

6.4 The Modified Criterion of Dispositionality/Categoricality

The adoption of ConD/C requires the revision of the @-criteria presented in Sect. 1.2. To illustrate the difficulty, let us first recall the criteria in question:

@-Criterion of Dispositionality The first-order state of affairs of an object instantiating a fundamental dispositional property is by itself (part of) a

minimal truthmaker for specific (perhaps, world-relative) modal truths (expressed by specific non-trivial counterfactuals) which concern the bestowal of specific (perhaps, world-relative) powers on the object.

@-Criterion of Categoricality The first-order state of affairs of an object instantiating a fundamental categorical property is not by itself (part of) a minimal truthmaker for specific (perhaps, world-relative) modal truths (expressed by specific non-trivial counterfactuals) which concern the bestowal of specific (perhaps, world-relative) powers on the object. In order to be (part of) a minimal truthmaker for the aforementioned truths, it must be supplemented with a nomic fact relating the property in question and other properties and/or relations.

Now consider the set of states of affairs consisting in the instantiations of some actual fundamental dispositional properties by objects which taken collectively will be sufficient for being a minimal truthmaker of a specific modal truth. Assume furthermore the almost unanimously accepted thesis of Truthmaker Necessitarianism (TN for short), according to which, if an entity e is a truthmaker for a truthbearer B, then, *necessarily*, if e exists then B is true.[19] In line with @-criterion of dispositionality, TN implies that, in *any* world in which the above-mentioned set of states of affairs exists, it must by itself be a minimal truthmaker for the specific modal truth. But according to ConD/C, fundamental properties are contingently dispositional. Hence, there must be a possible world in which at least some of the *actually* dispositional properties involved in the set are now categorical; and in *that* world the aforementioned set of states of affairs *cannot* by itself be a minimal truthmaker for the specific truth.

In order to address this difficulty, I suggest turning our attention to the state of affairs consisting of a fundamental property possessing the contingent higher-order feature of dispositionality (categoricality). Adapting Armstrong's well-known terminology, we may call such a state of affairs a *thick dispositional (categorical) property*. Now, in each possible world where an object instantiates a fundamental *thick* dispositional property, the associated state of affairs *must* by itself be (part of) a minimal

[19] Armstrong (2004, 6–7) presents an argument by *reductio* in favour of TN.

truthmaker for some specific truths. I suggest, then, the following as the proper ontological criteria for the dispositional/categorical distinction:

@^{T}-Criterion of Dispositionality The first-order state of affairs of an object instantiating a *thick* fundamental dispositional property is *by itself* (part of) a minimal truthmaker for specific (perhaps, world-relative) modal truths (expressed by specific non-trivial counterfactuals) which concern the ascription of specific (perhaps, world-relative) powers to the object.

@^{T}-Criterion of Categoricality The first-order state of affairs of an object instantiating a *thick* fundamental categorical property is *not* by itself (part of) a minimal truthmaker for specific (perhaps, world-relative) modal truths (expressed by specific non-trivial counterfactuals) which concern the ascription of specific (perhaps, world-relative) powers to the object. In order to be (part of) a minimal truthmaker for the aforementioned truths, it must be supplemented with a *nomic* fact relating the property in question and other properties and/or relations.

In the initial presentation of the truthmaking-criteria in Sect. 1.2 I did not discuss any objections to them. Here I would like to examine few worries related to their revised form. A first concern is that the above criteria promote a '*cheap*' metaphysical explanation of the fact that a fundamental property is dispositional (categorical). For some, the claim that the ground of the dispositionality (categoricality) of a fundamental property involves a higher-order ontological feature of dispositionality (categoricality) appears to be at least an *uninformative* metaphysical explanation. I do not, however, share this conviction. The adequacy of the suggested explanation via the posited feature is based on the ascription of a certain metaphysical *role* to the feature under consideration. As I have already explained, this role has to do with truthmaking. Ontological parsimony dictates that there actually is one ontological feature, P_D, shared by all dispositional properties, and similarly one ontological feature, P_C, shared by all categorical properties.[20]

[20] Of course, there are alternative options here. In my view, however, following either of these options leads to an unnecessary ontological proliferation with no obvious advantage.

Furthermore, the theoretical adequacy of $@^T$-criteria can be challenged in the following way. One presupposition of their implementation is that there are *some* specific (though perhaps world-relative) modal truths concerning the behaviour of the bearer of a dispositional property under various circumstances and expressed by non-trivial counterfactuals. The challenge is that the counterexamples to conditional analysis may be so interpreted as to suggest that there are *no* non-trivial counterfactuals made true by each fundamental property. If that is true, $@^T$-criteria collapse because both dispositional and categorical properties (more precisely, the states of affairs consisting in objects instantiating them) do not make true any specific modal truths.

One might respond to this challenge by arguing that there are no counterexamples at the fundamental level, and, therefore, the objector is not entitled to claim that there are no specific truths associated with each property.[21] An alternative response, which can work in all cases of finks, masks, and so on, would be to point out that the reaction of the objector to the presence of counterexamples is most probably exaggerated. For it is a reasonable claim that the existence of counterexamples to the conditional analysis of dispositions does not show that there are no counterfactuals (and related modal truths) associated with dispositional properties. Counterexamples merely show that the true counterfactuals in question are far more complicated than initially thought. The manifestation of a dispositional property for each of the various different situations it may find itself in corresponds to a distinct modal truth *and finks and antidotes are parts of some of those contingent circumstances.* Jacobs (2011) endorses such a view and articulates an account of the proper form of the counterfactuals which can express the relevant modal truths.[22] Following this

[21] Bird (2007, 60–63) makes a strong case for the absence of finks at the fundamental level.

[22] For Jacobs, the antecedent of such a counterfactual is a complex of property instantiations and the consequent is a singular causal relation holding between the property-complex appearing in the antecedent and some other property-complex which is the would-be effect. To address the difficulty from counterexamples, the property-complex appearing in the antecedent of a relevant counterfactual should either include or rule out various possible interference factors. Note, finally, that since in this case we attempt to find an ontological link between properties and counterfactuals, the charge of triviality of the truth of the latter is unfair. For, our goal is not to analyse a disposition of an object to M in circumstances S in terms of a counterfactual which ensures that that object would M in S *unless it does not.* It is rather to present by appropriate means the *complex* modal truths ontologically associated with a fundamental property of the object.

strategy, one may plausibly claim that in each possible world in which fundamental dispositional properties exist specific modal truths associated with them hold.

By embracing Jacobs' strategy we can also handle the problems arising from the possible existence of *intrinsic* finks and masks at all levels (not only at the fundamental one). This point is important, because the presence of intrinsic interference factors seems prima facie to undermine all solutions suggested by the advocates of conditional analysis and, furthermore, as Cross (2012a) points out, to cast doubt on our very intuitions about whether some objects have specific dispositions or not. Both consequences (if true) would be disastrous for the plausibility of the truthmaking account. For, the former would show that any optimism for the existence of an adequate solution to the various counterexamples is unfounded. While the latter would show that what an object would do in various conditions does not determine (or is totally irrelevant to) what dispositions it has. That in effect could be construed as undercutting the very justification of positing an ontological link between the ascription of properties and the truth of specific counterfactuals.[23] Fortunately for the truthmaking account neither of the above actually holds. First, Jacobs' account is able to take care of intrinsic finks and masks by including them (just as extrinsic ones) in the antecedent of the relevant counterfactuals. Second, it is not the case that what an intrinsically finked or masked object would do does not determine what dispositions it has. On the contrary, the object would do what it is supposed to do when finked or masked respectively. As I have remarked, objects have various dispositions to do specific things in various circumstances. The presence of intrinsic finks and masks delineates some of those circumstances (or even all of them, if it turns out that an intrinsic fink or mask is triggered by *any* stimulus whatsoever or is spontaneous). Hence, the object has specific dispositions even if it is intrinsically finked or masked. That would cause an epistemic difficulty in distinguishing an intrinsically finked disposi-

[23] In fact, if that were true, it would also undermine Ingthorsson's (2013) suggestion according to which some counterfactuals, though irrelevant to the essence of properties related to them, have a heuristic value for the identification of those properties.

tion from a non-present one; but, I think, it would cause no ontological trouble.

I would like to conclude this chapter by addressing another challenge which arises from a recent suggestion of Troy Cross. Cross (2012) presents an interesting view that blurs the ontological distinction between categorical and dispositional properties, while at the same time casting doubt on the basic idea which supports my preferred $@^T$-criteria. He starts off with the assumption that all philosophers may plausibly agree that categorical properties *can* endow powers, though *different* ones in *different* possible circumstances. So, for any categorical property F, there are possible conditions C (which, crucially, may include laws in other possible worlds) and a specific property G (associated with C) such that

In C, F disposes things to become G (6.1)

He then takes a further step and argues that it is a general feature of power and dispositional talk that the following equivalence holds:

In C, F disposes things to become G iff F disposes things to become G in C (*)

Hence, asserting (6.1) is tantamount to asserting

F disposes things to become G in C (6.2)

Cross rightly points out that (6.1) and (6.2) differ in modal strength. The former ascribes a *contingent* conditional power to F (a power that F has only in worlds where C holds), while the latter ascribes a *necessary* conditional power to F (a power that F has in *all* worlds, regardless of whether C holds or not). In spite of that difference, however, he insists that, due to (*), we are entitled to move from (6.1) to (6.2). The last step of Cross' argument is to interpret (*) as an expression of a general *ontological* feature of powers, according to which all potential differences in powers a property bestows on its bearers are *also* actual differences in potential. On this basis one may arguably show that *both* dispositional and categorical properties necessarily bestow powers on their bearers under actual and non-actual activating conditions, and so the ontological distinction between the two types of property is blurred.

As far as I can see, the last conclusion should be resisted, because Cross' suggested analogy can be sustained only on pain of begging the question against the distinction between categorical and dispositional properties as traditionally conceived. In order to show that, we have to point out that (6.1) and (6.2) do not only differ in their modal strength. They also (and crucially) differ in that it is only by following (6.1) that we can ascribe a *genuinely relational* power to F; a power that can be based on a relation between F and C which *both* exist in the *same* world. On the contrary, by following (6.2) we cannot ascribe a *genuinely relational* power to F because, according to (6.2), F can also have the power in question in worlds where F, *but not C,* exists. *This* difference, however, shows that in the case of categorical properties one cannot pass from (6.1) to (6.2), unless, of course, one rejects what has been almost unanimously accepted as the crucial characteristic of categorical features; namely, that they cannot *by themselves* bestow any powers on their bearers. One might also challenge Cross' attempt to blur the categorical/dispositional distinction by refuting his inclusion of laws of nature in the set of possible activating conditions of powers. Perhaps, Lewisian laws, *qua* supervenient on the instantiations of natural properties, can be construed as 'globalised' forms of the usual 'local' stimuli triggering the manifestation of properties. But what grounds do we have to extend, for instance, the analogy to the case of laws construed as relations between universals? I conclude that Cross' proposal poses no threat to my suggested $@^{T}$-criteria for the ontological distinction between fundamental dispositional and categorical properties.

7

Do Nomic Relations Exist?

Thus far I have argued for a metaphysics of fundamental natural properties and relations of the actual world. In particular, I put forward the thesis that *fundamentally* our world is a categorical–monistic one (henceforth, a C-world). A metaphysics of properties almost always comes together with a corresponding metaphysics of laws of nature. This is obviously true in the case of laws which essentially involve ontologically robust natural (fundamental) properties and relations. Instances of that case is the DTA[1] theory of laws as relations between universals and, more generally, any account of laws as relations between properties construed as belonging to a distinct (from concrete particulars) ontological category. But it is also true even within broadly nominalistic metaphysical contexts where we do not have a substantial notion of a natural property. The reason, in that case, is that one of the main roles of natural properties is to help explain the behaviour of their bearers. It is assumed, however, that laws of nature, if they exist, purport to fill the same role; hence, the inevitable connection. So, in the chapters to follow, I'll present the basic metaphysi-

[1] This is an acronym for the theories of laws as relations between universals presented (independently) by Dretske (1977), Tooley (1977) and Armstrong (1983).

V. Livanios, *Science in Metaphysics*,
DOI 10.1007/978-3-319-41291-7_7

cal consequences of my preferred account of the nature of fundamental properties for laws of nature.

A considerable part of the remaining discussion in this book concerns the plausibility of a specific view about laws, the so-called Dispositional Essentialist Account of Laws (DEAL). DEAL emerges naturally from the core principles of Dispositional Realism in its orthodox form, and its critical examination is an important issue both for my purposes and for the general debate concerning the metaphysical status of laws. First, the acceptance of DEAL has important consequences for the *ontology* of laws, because it either threatens their *genuine* existence or implies their non-existence *tout court*. In particular, according to a *reductionist* version of DEAL (Bird 2007, Ch.9), laws *supervene* on the dispositional natures of fundamental properties and relations. This interpretation, backed up by a specific notion of supervenience in which the supervenient entities are no genuine ontological additions, may easily threaten the genuine existence of laws, ascribing to them at best an ontologically inferior status (for instance, laws may simply summarise the 'behaviour' that properties bestow on their bearers according to their natures). Things are worse for laws according to an *eliminativist* version of DEAL (Mumford 2004); from that perspective, laws have no ontological work to do and simply do not exist. Second, the proponents of DEAL reject what I think is one of the main characteristic features of laws of nature, their metaphysical contingency. A robust form of the latter is a crucial consequence of ConD/C, an account which preserves the basic intuition behind the notion of laws (a fact that advocates of DEAL also acknowledge but nevertheless try to dispel because it does not conform to their theory). Hence, showing that DEAL is implausible removes an essential obstacle to accepting ConD/C, since the former rejects one of the core consequences of the latter. Finally, and more importantly for the general debate, DEAL is steadily gaining popularity nowadays, tending to become the new orthodoxy in the metaphysics of laws of nature. Fresh arguments against it surely contribute to one of the liveliest debates in contemporary metaphysics.

Obviously, the first question we need to address is, Do laws of nature exist? In order to provide an answer to this question, I must first make clear the ontological view of laws of nature I presuppose in the following discussion. In common with a number of property theorists, I endorse the view that laws of nature express (one way or the other) relations that relate fundamental natural properties and relations. This remark is vague as it stands, but one way to elucidate its meaning (the one I prefer) is to construe laws of nature as second-order facts (or, following Armstrong (1997), states of affairs) which 'emerge' when the nomic relations of the actual world relate their relata. More precisely, a law of nature is a fact of the form F(P, Q, R,…), where F is a most often (but, in my view, not necessarily so) functional nomic relation and P, Q, R, and so on are its relata, which, as I have already noted, are fundamental natural properties and relations. I also take it for granted that law-facts cannot exist in a possible world unless *all* of their ontological constituents exist in that world too. In what follows, then, I'll presuppose the existence of all properties-relata and concentrate only on nomic relations. Given all that, the question concerning the existence of laws of nature boils down to the question concerning the existence of nomic relations.

7.1 On the Ontological Status of Nomic Relations

The answer to the question appearing in the title of this chapter would be easy, had *all* possible worlds been like the actual, categorical–monistic one. Recall that, according to Categorical Monism *all* actual fundamental natural properties and relations are categorical. Fundamental categorical properties and relations, however, have a nature which is metaphysically unrelated to their nomic/causal roles. Hence, according to Categorical Monism at least those relations (between properties) that express the *nomic* roles of their relata (i.e., the nomic relations) cannot supervene on the nature of the relata. This feature ensures that for categorical monists *all* nomic relations are by definition *external*.[2] Noticeable examples are Armstrong's (1983) account of laws as states of affairs expressing external

[2] Armstrong (2004a) proposed an account of instantiation that under certain assumptions may lead to the conclusion that nomic relations with *any* kind of relata should be internal. I present my reasons to refute this account—at least for the case of nomic relations—in Sect. 8.1.1.

relations of nomic necessitation between universals, and all those sophisticated regularity theories about laws (like the so-called Mill–Ramsey–Lewis [MRL] account) according to which, roughly, natural laws express external relations of extensional inclusion between natural properties. In both cases, laws determine, in each possible world in which they exist, the (nomic/causal) role of each of their relata in that world. The crucial point for the issue under consideration is that the metaphysical element of non-supervenience that constitutes (by definition) the external character warrants the almost unanimous agreement between metaphysicians that, if external relations exist at all, they *genuinely* exist as basic constituents of the ontological furniture of the world. As a natural consequence, categorical monists must admit that nomic relations belong to the ontological furniture of any possible world that is correctly described by their theory. And, of course, for them the actual world is a C-world.[3]

Given that the actual world is a C-world, the above remarks indicate that the actual nomic relations are genuine existents. It might be objected, however, that though in a C-world there exist external relations between fundamental properties, these relations *cannot be nomic*. Any genuine existing relation which has the credentials to be nomic has to play a *distinctive* role in the world in which it exists. One therefore might challenge the nomic character of a relation between properties by arguing that the relation in question does not (or cannot) play that role. Of course, the objection is empty unless the objector determines what exactly this nomic role is. There is no consensus on that issue, since different philosophers propose different things a law (and its associated nomic relation) must do. The rival views fall under two general perspectives. From the first, 'light-weight', perspective, laws have no governing role at all and simply *describe* either the regularity patterns of non-modal facts already existing in a world or the essences of fundamental dispositional properties. From the second, more 'heavy-weight' perspective, genuine laws should have a more *robust* role in the world in which they exist, usually in the sense of *determining* the course of events occurring in it. I am sympathetic

[3] The fact that, in a C-world, all nomic relations are external does not preclude the possibility that other, non-nomic, relations between natural properties are internal.

to the second view (though I cannot argue for my choice here); that is why I think that eventually DEAL is not a metaphysical account of *genuine laws* at all. In other words, if in a possible world DEAL holds, there are no *genuinely* existing nomic relations in that world. The same, I think, would hold in a C-world if the Humean MRL account were true. The external relations *describing* the regularity patterns are not genuine nomic relations either. In my view, the only genuine nomic relations in a C-world must be a *kind* of DTA external relations between fundamental categorical properties, from which we can 'get' (in the sense that nomic relations determine them) the causal/nomic roles of properties and the course of events in the world in question. (Notice that, from that perspective, nomic relations have modal power; a fact which, however, does not challenge the categorical character of a C-world, since that character concerns only first-order fundamental properties and relations.)

Returning now to the actual, C-world, how can external nomic relations play their distinctive nomic role? The question is pressing, because in the absence of a reasonable answer to it, defending nomic realism makes no sense. And, as I remarked above, there exist arguments purporting to show that there *cannot* be such an answer. In the sequel, I'll sketch my response to that claim focusing on Mumford's (2004) arguments against nomic realism. More precisely, I'll examine one of the horns of his Central Dilemma, the one allegedly showing that external relations between fundamental properties cannot play any 'governing' role in the actual world. For Mumford, this 'governing' role consists in the determination of properties' causal/nomic role and the course of world events (*ibid.*, 146). The first aim can in principle be achieved, provided that we have the totality of external nomic relations among actual fundamental properties. In that case, the only thing we have to do is to construct the Ramseyfied lawbook of the actual world and follow Hawthorne's (2001) instructions to determine the causal/nomic role of any fundamental property we choose (for a brief description of the construction, see Sect. 4.1).[4] In his argument, Mumford concentrates on the aim of determining

[4] No doubt, *epistemological* problems beset that procedure, but, as far as the *metaphysics* is concerned, it is a perfectly legitimate way to determine the causal/nomic role of any natural fundamental property.

the course of events. His main concern is to show that nomic relations and laws which are not 'in' the beings of their instances cannot determine what actually happens. Crucially, he also seems to think that, in this case, being 'in' should be interpreted as being *spatiotemporally* 'in'. This belief leads naturally to the conclusion that the only remaining option for the advocate of governing laws is to endorse Armstrong's 'Aristotelian' interpretation of nomic realism where laws, *qua immanent* universals, 'live' spatiotemporally in their instances. (The latter are the particular sequences of events which involve the instantiated universals associated with the nomically related fundamental properties.) This is not true, however, since laws *can* be *non-spatiotemporally* 'in' their instances. I'll soon come back to this point but, for the time being, let us skip it and discuss Mumford's objections to Armstrong's externalised account of laws.

In Armstrong's view, there is nothing more to the 'governing' relation between laws and events than the usual *exemplification* relation holding between a universal and its instances. Prima facie then there is no problem for Armstrong's external nomic relations to govern actual events. However, a difficulty pops up related to the Instantiation Condition that Armstrong's naturalistic approach presupposes. According to that condition, as Mumford (*ibid.*, 149) points out, it is a law that N(F, G) only if some particular x instantiates F (and thereby G) at some instant of time t. A question then seems to arise: How can laws determine their instances when those very instances seem to be metaphysically prior to laws (in the sense that they are required in order for the laws to exist in the first place)?[5] Here I think that Mumford misidentifies the real source of the problem. The Instantiation Condition requires just *one* instantiation of F and G in the entire spatiotemporal history of the actual world in order for the law N(F, G) to exist as universal. As far as I can see, however, this assumption by itself cannot justify the claim that the *entire* pattern of instantiations of the law in question is *metaphysically prior* to the law itself. Hence, I think that the Instantiation Condition leaves room for the law to determine the vast majority of events, since its existence is de facto ensured by just one of them. The real problem pointed out by Mumford

[5] For a detailed exposition of the same problem, though expressed in causal terms, see also Bolender (2006).

is associated with the metaphysical account of the relation of exemplification that Armstrong embraces. It is well known that Armstrong has suggested that our world is fundamentally a world of states of affairs; universals and thin particulars are nothing but *abstractions* from these basic building blocks of reality. From that perspective, there is no need for a metaphysically *robust* relation of exemplification acting as metaphysical 'glue'; the unity of (thin) particulars and universals is guaranteed since they both exist only as abstractions from states of affairs. It is *that* account (extended to cover law-universals too) which creates the 'metaphysical priority' problem, for Armstrongian laws seem to govern the very relational states of affairs which they are ontologically dependent on. If my diagnosis is correct, one may then avoid the difficulty not by rejecting the claim that governing amounts to a kind of exemplification, but by suggesting instead a different metaphysical account of the exemplification itself. In fact, a moderate naturalist[6] might think of the exemplification relation as a primitive, *non-spatiotemporal* metaphysically robust relation between universals and their instances. Extending this account of exemplification to the external nomic relations-universals, she might have a conception of nomic governing (as exemplification) not facing the above-mentioned ontological dependency difficulty. For, there is no reason, according to that account, to think of external nomic relations-universals as ontologically dependent on their instances. In any case, Mumford himself leaves the issue open when he claims that there might be another theory which treats external relations as nomic ones *and* vindicates their governing role. For him, the decisive difficulty besetting any kind of externalised laws is the implied quidditism. External nomic relations allow laws and (nomically related) properties to vary independently and this leads to the 'unpalatable' view of quidditism. In response, I refer the reader to Sect. 4.3, where I defend RNDV against the major objection to it (the one related to the so-called Permutation Difficulty).

Now, were the whole issue to be discussed in this section the genuine existence of actual nomic relations (or, more generally, of C-worlds'

[6] Moderate, because she only has to admit that the exemplification relation is a metaphysically robust, primitive non-spatiotemporal entity. For an account akin to this but in a non-naturalistic context where universals themselves are abstract, non-spatiotemporal entities, see Moreland (2001).

nomic relations), I would have happily concluded the section here by answering the question posed in the title in the affirmative. Recall however, that Categorical Monism, if true, is *contingently* true. Hence, in other possible worlds (even granting that they are populated by counterparts of actual properties) either Dispositional Monism or Dualism is true. Consequently, in order to answer the general existential question concerning nomic relations, we have to examine their ontological status in dispositional-monistic and property-dualistic worlds.

Prima facie, the issue concerning the genuine existence of actual nomic relations would also be clear provided the actual world had been as dispositional monists describe it. Recall that dispositional monists claim that *all* actual fundamental natural properties and relations are dispositional. Most of them are essentialists and claim that all genuinely dispositional properties have an intrinsic nature (and an identity) that is exhausted in their essential causal roles, or, in other words, in the causal powers which they essentially bestow on their bearers.[7] According to dispositional essentialists, the intrinsic dispositional nature of natural properties (and relations) determines their causal/nomic roles in each possible world in which they exist, and so it also determines *completely* and *necessarily* the instantiation of nomic relations that laws involve.[8] This latter belief (but not *only* this, as the case of Mumford (2004) who is not a dispositional essentialist clearly shows) leads either to nomic *eliminativism* or to DEAL. Recall that, according to DEAL, nomic relations just 'express' the dispositional essences of fundamental properties and relations. Those dispositional monists who do not want to eliminate laws from the ontological landscape claim that nomic relations *supervene* on the natures of their relata. If we follow Lewis' popular definition (1986, 62), what dispositional monists actually claim is that all nomic

[7] The criterion of dispositionality for fundamental properties of the actual world does not have to be related (as dispositional essentialists claim) to the essentialist claim that causal/nomic roles are essential to properties. Hence, all those philosophers who accept the existence of genuine, irreducible dispositional properties do not have to follow the essentialist course.

[8] Of course, this only holds under the assumption that fundamental properties and relations are dispositional in all worlds in which they exist. Anyone (like me) who rejects this assumption must restrict the range of validity of the claim to those worlds in which properties retain their dispositional character.

relations are *internal* relations between fundamental natural properties and relations.[9]

The metaphysical element of supervenience which constitutes (by definition) the internal character has led the majority of metaphysicians to ascribe an ontologically *inferior* status to internal relations. For instance, David Armstrong is well known for upholding the thesis that what supervenes is not something ontologically more than what it supervenes on (in his words, the former is an 'ontological free lunch').[10] Hence, for Armstrong, given that internal relations do supervene on the natures of their relata, they are not a genuine ontological addition to the world's furniture (Armstrong 1997, 87).[11] The point can be vividly illustrated in terms of truthmakers. As Simons (2010) points out, in the case of an internally true relational predication, there is no need to suppose the existence of a *relational* truthmaker, since what makes true the predicative sentence is either the relata themselves or some of their essential properties constituting their nature.

> It therefore appears generally to be the case that from an ontological point of view there are no such things as internal relations. (Simons 2010, 205)

Given the previous points concerning the ontological inferiority of internal relations, dispositional monists must admit that, in any possible world that is correctly described by their theory (a D-world, henceforth), nomic relations should be downgraded to ontologically 'second

[9] There are at least two different senses of internality. According to the weak sense, internal relations are grounded merely in the numerical identity of their relata, while according to the strong one, they are grounded in the qualitative natures of the relata. In the following discussion, I presuppose the (more popular) strong sense of internality.

[10] Armstrong uses a definition of supervenience that does not refer to groups of properties. For him, an entity Q supervenes on entity P iff it is impossible that P could exist and Q not exist, where P is possible (1997, 11). I prefer the more or less standard definition, according to which internal relations supervene on the natures of their relata iff there cannot be a difference in the relations without a difference in the natures of their relata.

[11] In his earlier work (1978a, 86), Armstrong presents two reasons in favour of the thesis. First, according to his a posteriori realism, it is a suspicious fact that one can discover the existence of internal relations simply by knowing the properties of their relata. Second, it seems implausible to think that internal relations genuinely exist, given that we have no reason to attribute causal efficacy to them. In his subsequent work, however, Armstrong acknowledges that 'it is not clear how the thesis that what supervenes is no addition of being is to be proved' (1997, 12).

grade' entities, which at best supervene on the (dispositional) natures of fundamental natural properties (and relations). Hence, it seems that the ontologically genuine existence of *all* nomic relations is threatened in D-worlds. And of course, for dispositional monists the actual world is a D-world.[12]

Let us now turn to property-dualistic worlds (or M(ixed)-worlds, as I call them). Embracing Property-Dualism seems to make things easier for the nomic realist,[13] because, if a possible world w is not a D-world, there is no prima facie reason to doubt the genuine existence of *all* nomic relations of w. In principle, w, *qua* M-world, may have three distinct types of nomic relations. The first type includes all nomic relations with only categorical relata. By contrast, the second type includes all nomic relations with only dispositional relata. Finally, the third type includes all 'mixed' nomic relations relating both categorical and dispositional relata. As I have already remarked, and regardless of the ontological status of mixed nomic relations,[14] members of the first type genuinely exist. Notice, however, that the existence of both kinds of fundamental features in M-worlds does not by itself imply that nomic relations of *all* types exist. For, there is always a possibility that all nomic relations of an M-world have only *one* kind of properties as their relata. If that kind happens to be the second one, nomic realists find themselves in the same predicament as in D-worlds. In other words, in that unfortunate for the nomic realist case, an M-world (just like a D-world) has ontologically 'second grade' nomic relations which at best supervene on the (dispositional) natures of fundamental natural properties (and relations).

Nomic realism in M-worlds is not only threatened by the above possibility. To see that, consider the view which some property-dualists adopt for actual nomic relations. It is a version of DEAL which 'absorbs' the nomically related *categorical* features appearing in mixed nomic relations into the essences of the dispositional relata (see, for instance, Ellis 2001).

---

[12] As in the case of the C-world, the fact that, in a D-world, all nomic relations are internal does not preclude the possibility that other, non-nomic, relations between natural properties are external.

[13] The expression 'nomic realist', as I use it here, refers to those philosophers who acknowledge the genuine existence of at least some nomic relations in some possible worlds.

[14] It is not clear whether mixed nomic relations are external or internal and, consequently, whether they genuinely exist or not (more on this issue in Sects. 8.2 and 8.3).

Assume now that in an M-world w there are no nomic relations of the first type. To the extent, then, that the above-mentioned version of DEAL *could be* true for all types of nomic relations of w, the genuine existence of *all* nomic relations can be threatened even in M-worlds.

I am a nomic realist and so I would like to defend here the genuine existence of (at least) some nomic relations in M-worlds. To this end, and given that the appeal to the 'absorbing' version of DEAL in an M-world creates the worst scenario for the nomic realism I intend to defend, I'll challenge DEAL. Crucially, my task is not only to argue that the 'absorbing' version of DEAL is not tenable; it is also to show that even some nomic relations of the *second* type cannot be plausibly construed as 'flowing' from the dispositional essences of their relata (and, consequently, that it is not so clear that they are internal, ontologically second-rate, relations). The two cases to be discussed in the sequel are not novel to the debate between property-dualists and property-monists. Both of them are mentioned by Bird in the concluding chapter of Bird (2007) as potential problems for DEAL. The first one is related to symmetries and conservation laws and will be examined in Sect. 7.2, while the second concerns all nomic relations involving fundamental constants and will be studied in Sect. 7.3. The aim in both sections is to show that, *pace* Bird, the above-mentioned cases raise serious difficulties that DEAL cannot meet.

7.2 Against DEAL: The Case of Symmetries and Conservation Laws

It is a fact about our world that certain quantities are conserved in all interactions. Examples of such quantities are mass–energy, electric charge, momentum, angular momentum, and so on. It is also well known that contemporary physical science strongly suggests that specific kinds of symmetries (the continuous ones) are intimately related to the conservation of these physical quantities. In particular, a famous theorem of Amalie Emmy Noether (1918) says that for each continuous symmetry of the Lagrangian function of a physical system there is a quantity which is conserved by its dynamics.[15] For instance, invariance under time translation implies conservation of the system's energy, invariance under translation in space implies conservation of its momentum and invariance under spatial rotations implies conservation of its angular momentum. The application of Noether's (first) theorem in the case of various continuous symmetries provides a *unified* and *non-ad hoc* explanation of the existence of conservation laws and conserved quantities in our world.[16] It is a unified explanation, because the theorem provides a unique procedure through which each conserved quantity emerges. Just take the right kind of symmetry and the theorem will yield the corresponding conserved quantity as a necessary consequence. Furthermore, it is not ad hoc because the explanation is grounded in the concept of symmetry which has a broad explicative role in physical science in areas quite irrelevant to the issue of conserved quantities. Hence, the remarkable fact of the conservation of some fundamental properties is explained by a procedure showing that it is a consequence of the invariance under the actions of antecedently well-understood symmetry transformations (like translations in space–time, spatial rotation, etc.).[17]

Now, the crucial question is this: Is there a version of DEAL capable of accommodating the case of conservation laws? In what follows, I explore a course dispositional essentialists have followed in order to offer an adequate explanation of the existence of conservation laws. The suggested explanation is an ontological one and it comes in two versions. According to the first version (Bigelow et al. (1992)), our world is the unique actual member of a natural kind the real essence of which requires conservation of certain quantities (in other words, conservation

---

[15] I refer here to the so-called *variational* symmetries that leave the form of the Lagrangian invariant up to a divergence term. Every such symmetry is also symmetry of the Euler–Lagrange equations that follow from the application of the action principle. However, the converse is not always true.

[16] The symmetry-based explanation is not necessarily an *ontological* explanation. Of course, it can be one provided that fundamental symmetries are properties (not necessarily essential) of the world (or, of its structure).

[17] That is not true for the conservation of various *charges* which is a consequence of invariance under *internal* symmetry transformations (related to abstract spaces). So in this case we do not have an intuitively well-understood concept to ground the explanation. Nevertheless, I think it is an adequate explanation because it is an instance of (an extension of) an explanatory procedure which is broadly successful.

laws emanate from the kind-essence of the actual world). Bird, a dispositional essentialist who does not acknowledge the ontologically robust existence of natural kinds, suggests (2007, 213–4) the second version of the above ontological explanation. Since for him laws of nature 'flow' from the essences of fundamental natural properties (and not from kind-essences), he suggests (without finally invoking) that conservation laws possibly emanate from the dispositional essence of the property 'being a world (like ours?)' which the actual world definitely instantiates. Despite their initial appeal, I think that both suggestions can be characterised as either ad hoc or poor explanations. Take, for instance, the property-essence version. The property 'being a world (like ours)' is either posited only in order to explain conservation facts (so, in that case, we have an ad hoc explanation) or, due to its 'global' character, is meant to provide an explanation of *all* worldly facts (so, in that case, we have a very poor and extremely coarse-grained explanation. Consider, analogously, the property 'being a man [like Socrates]' that purports to explain all Socrates' traits).

The suggestion of Bigelow et al. is more elaborate. They start interpreting laws of nature as arising out of essences of particular natural kinds (of substances, events and processes). They soon observe, however, that conservation laws differ from all other laws only in scope; that is, they do not apply to particular kinds of events and processes, but to all kinds (*ibid*., 384–385). Furthermore, they notice that conservation laws apply non-approximately only to events and processes occurring in closed and isolated systems, and that, strictly speaking, it is only the universe itself that is perfectly closed and isolated. So, they introduce the world-as-one-of-a-kind view as a natural extension of their initial view. In order not to be accused of offering an *ad hoc* solution to the problem of conservation laws, Bigelow et al. use the world-as-one-of-a-kind view to provide a covering explanation for the emergence of *all* laws of nature. They claim that, if conservation laws arise out of the essence of the whole world, and granting that conservation laws differ from the other ones only in scope, then other laws (concerning parts of the world) must also arise out of the world-essence. Moreover, they suggest that the essences of parts of the world, and the correlations which depend on them, may both contribute to the world-essence. And only insofar as they do that, laws of nature can arise out of them (*ibid.*, 386–7). I am not confident that this last move saves Bigelow et al.'s suggestion from the accusation of ad hoc-ness. But, surely, the world-essence explanation is at least *redundant* as far as the case of non-conservation laws is concerned. Since Bigelow et al. have already provided a fine-grained explanation of non-conservation laws as emanating from particular kind-essences, what more (besides, of course, escaping the ad hoc-ness charge) can the global, coarse-grained explanation in terms of the world-essence offer?[18]

I will not discuss any further the details of the above proposal, because I think there is another line advocates of DEAL may follow, which dissolves the problem arising from the existence of conservation laws. The crucial move is to shift the discussion from conservation laws to symmetry principles and regard the latter as *alternative formulations* of the former. As far as symmetry principles themselves are concerned, a friend of DEAL (following Bigelow et al. 1992, 371) may claim that they are concerned with the essential properties of the kind of world we live in; they are derived from the fact that our world (and all worlds belonging to the same natural kind as ours) *essentially* displays certain physical symmetries. Though this alternative explanatory move is not explanatorily superior to the preceding one (we have, once again, a kind of ontological explanation of physical facts based on displays of the world-essence), it affects the symmetry-based explanatory proposal I have put forward in a way that the latter does not. For if symmetry principles are just alternative formulations of conservation laws, they cannot be used to ground genuine explanations for the existence of the latter. Rather, the world-essence explains all worldly facts, among which are facts describable either via conservation laws or via symmetry principles.

---

[18] Alan Chalmers (1999, 14) raises two more worries about Bigelow et al.'s suggestion. He points out, first, that conservation laws—just like the other (non-cosmological) laws—are applied to *local* systems in the world, not to the world as a whole. Though he admits that conservation laws apply exactly only to *isolated* systems, he does not regard this as a cogent reason to suppose that those laws apply only to the whole universe; for all laws apply exactly only in idealised contexts (in his words, 'there is no causal factor governed by any law which operates totally unhindered'). The second worry is that, by appealing to the world-essence, Bigelow et al. do not eventually succeed in offering a single *unified* account of laws of nature. For, while conservation laws impose (often strong) constraints on the physical processes in the universe, they do not 'describe powers and capacities that are exercised in a law-like way in bringing about the phenomena of the world as causal laws do'.

How plausible, however, is the thesis that symmetry principles are just alternative formulations of conservation laws? There is certainly a connection between the two (due to Noether's first theorem), but this fact by itself does not provide evidence about their equivalence. First, the connection between symmetries and conservation laws is confined only to continuous variational symmetries. Second, even in the case of continuous symmetries, the connection is only established provided that the laws of motion are in the form of differential equations derivable from an action (Hamilton's) principle.[19] Although the majority of fundamental laws of motion in physics satisfy this condition, it has not been proved that this holds for all physically interesting cases. Given our ignorance about the necessary and sufficient conditions characterising those systems of differential equations which are equivalent to a system of Euler–Lagrange equations,[20] it is at least premature to say that the aforesaid fact represents a feature of our world rather than a basic methodological assumption for the construction of physical theories.[21] Third, symmetry principles are conceptually related to (and constitute an explanatory basis for) many issues that are entirely independent from conservation facts. So, the alleged equivalence with conservation laws seriously downgrades their significant role. Hence, to recapitulate, symmetry-based considerations provide a unified and non-ad hoc explanation of the existence of conservation laws and conserved physical quantities. By contrast, DEAL yields poor or ad hoc explanations. Furthermore, the attempt to undermine the traction of symmetry-based explanations in favour of DEAL is grounded in the unwarranted hypothesis of the equivalence of symmetry principles with conservation laws.

Finally, it might be objected that the DEAL-friendly explanation is superior to the symmetry-based one because, unlike the latter, it is an *ontological* explanation. It is true that the symmetry-based explanation

---

[19] According to Hamilton's principle, the action integral $I = \int_{t_0}^{t_1} L \cdot dt$ of the Lagrangian function of a physical system is stationary under arbitrary variations dq_i of the generalised co-ordinates which vanish at the limits of integration.

[20] The application of Hamilton's principle yields the Euler–Lagrange equations. They are in general second-order differential equations for the Lagrangian function.

[21] For more details about that, see Earman (2004).

is not *necessarily* an ontological explanation, but, nonetheless, it can be one, provided that fundamental symmetries are natural properties (not necessarily essential) of the world (or, of its structure). And, as far as I can see, there is nothing preventing, *even an advocate of DEAL,* from embracing the ontological interpretation of fundamental symmetries.[22] Bird (2007, 214) disagrees; he argues that a proponent of DEAL should favour the *epistemic* interpretation, insisting that symmetry principles are *meta-statements* about laws of nature and not genuine laws of nature themselves (i.e., laws about the laws). His line of thought is the following. Since (in accordance with the orthodox version of DEAL) laws of nature are necessary, there is no need for any *further* constraints imposed on physical properties. Properties are necessarily constrained by their dispositional essences and so there is no need for any higher-order properties to determine which relations they can engage in. The upshot is that, first impressions aside, symmetry principles are pseudo-laws. They do not constitute a genuine addition to the ontological furniture of the world; rather, they ought to be regarded as (necessary?) features of the human way of representing the world.

Bird's claim would be convincing provided that the sole role of symmetry principles was to impose constraints on physical properties. However, that is not actually the case; the application of symmetry principles has a much wider scope, and this is a fact which in some cases (consider, for instance, the role of symmetry principles in the prediction of previously unknown elementary particles) grounds powerful arguments in favour of the ontological interpretation of symmetries. Hence, the plausibility of Bird's view heavily depends on the existence of independent arguments against the ontological viewpoint. It is not enough just to remark that the epistemic view about symmetries is *consistent* with DEAL (and with Dispositional Essentialism in general).

A friend of DEAL may turn the tables here and, *pace* Bird, claim that the adoption of the ontological viewpoint about symmetries may even benefit her project. For it might turn out that DEAL is a consequence of a more general theory based on fundamental symmetries of nature. In

[22] Recall (from Sect. 2.2.2) the two main approaches concerning the interpretation of fundamental symmetries, the ontological and the epistemic.

that case, an ontological interpretation of symmetries could be construed as extending (or refining, in some sense) DEAL. This suggestion, however, is based on a mere speculation about the *future* development both of physical theories and of DEAL. What I hope this section shows is that, at the current stage, an ontological interpretation of symmetries cannot be construed as refining or generalising DEAL (as we know it).[23] Hence I conclude that the case of conservation laws raises serious difficulties for DEAL, and that it is at least doubtful that an advocate of this account of laws has the means to meet them.

7.3 Against DEAL: The 'Constant' Threat

7.3.1 Introduction

In this section I focus on another difficulty for DEAL: the problem of fundamental constants.[24] Contemplating the existence and role of fundamental constants is important for both scientific and philosophical reasons. First, it has been shown that, within the scientific context, fundamental constants are intimately related to a variety of fundamental physical facts in need of explanation. For instance, the size and structure of almost all composite objects of our universe (from nuclei to stars, and beyond stars to galaxies and galaxy clusters) are manifestations of the various possible equilibrium states between competing forces of nature, the structure of which (i.e., states) is largely determined by fundamental constants (for a lucid explanation, see Barrow and Tipler (1986)). Furthermore, fundamental features of the universe itself, such as its

[23] French (forthcoming) is a recent investigation of the prospects of orthodox dispositionalism as a metaphysical account of the nature of fundamental symmetries. He examines various ways of implementing standard stimulus-manifestation dispositionalism in the context of physical symmetries and finds them problematic.

[24] The two difficulties (the one related to conservation laws discussed in the preceding section and the one associated with fundamental constants to be examined here) do not exhaust the problems that DEAL should meet. There is also a problem related to the least action principles (for a discussion, see Katzav (2004) and Ellis (2005a)) and another difficulty concerning properties (such as inertial and gravitational mass) which, though they seem to be involved in laws in accordance with the dispositional conception, are nomically related by a law that most probably is not an expression of their dispositional natures. All these issues are briefly discussed in Bird (2007).

rate of expansion, the presence of galaxies within it and, crucially, the existence of life are dependent on the values of certain fundamental constants (see, for details, the six numbers that govern the universe according to Martin Rees (1999)). As a consequence of the above facts, one may reasonably claim that we have real advances in our scientific understanding of the physical world when we learn more about fundamental constants. Barrow (2002, 61–66), for instance, lists six kinds of discovery which, each in its own way, increase our understanding of fundamental features of the actual universe. Among them is the enhancement of the status of a previously known constant (e.g., the velocity of light in the Special Theory of Relativity as the maximum velocity) and the explanation of the value of a constant from first principles of a physical theory.[25] Second, the existence of fundamental constants has also raised important issues lying at the border between science and philosophy. Undeniably, the most discussed topic related to fundamental constants is their alleged fine-tuning which seems necessary for the existence of the aforementioned features; this fine-tuning has become the principal motivation for the development of various explanatory schemes, ranging from anthropic arguments to a beneficent Creator who created the universe with the intention of producing human beings, and to the multiverse hypothesis. Nevertheless, I'll limit the discussion to the difficulty the existence of constants raises for DEAL.

Bird (2007, 211–12) briefly discusses the case of physical fundamental constants and presents an argument that articulates an objection to DEAL. Here is a reconstruction of his argument:

(1) Some laws of nature involve fundamental constants.
(2) There is a possible world in which the values of these constants are slightly different.

[25] Of course, extremely important is also the discovery of new fundamental constants that always accompany the introduction of novel fundamental theories (for instance, Planck's constant h in quantum theory and string tension λ in String Theory). Barrow's list is supplemented with the discovery that the value of one constant is determined by the values of others, the discovery that a physical phenomenon is governed by a new combination of constants, and, finally, the discovery that a quantity believed to be a constant is not really constant. For more details and examples, see Barrow (2002).

(3) Small differences in the values of constants would not require that the properties appearing in the laws of that possible world differ from the actual ones.
∴ We have different laws involving the same fundamental properties.[26]

A corollary of the above argument, if it is sound, is that the dispositional essence of fundamental, nomically related, properties and relations cannot account for the difference between the laws of the actual and the corresponding laws of the possible world. Consequently, DEAL is at best incomplete.

7.3.2 Objections to the Rescue?

Given the validity of the argument, the only escape route for the advocate of DEAL is to challenge the truth of at least one of its premises. In what follows, I'll consider objections supported (at least prima facie) by the findings and practice of contemporary physical science (or, at least, compatible with them). I strongly believe that, especially due to the nature of the issue under consideration, contemporary physical science must inform our metaphysical investigations on that matter. Having this in mind, I now proceed to the examination of the objections.

7.3.2.1 Objection to Premise (1)

It might be objected that premise (1) is false because:

> *There are no* ***fundamental*** *constants and consequently there are no laws of nature that involve fundamental constants.*

Bird thinks that this objection is sufficient to disarm the argument against DEAL. Yet, in order to be able to do that, it must at least be indepen-

[26] An implicit assumption of the argument is that the laws of the possible world in premise (2) differ from the corresponding actual laws. This assumption, however, is not controversial because if the constants appearing in the laws assume different values, then the propositional contents of the laws are surely different and that means (on any tenable account of laws of nature) that laws themselves have changed.

dently defensible. To evaluate its plausibility, we must first consider the definition of fundamentality for constants of nature and then examine the reasons we have to think that there are probably no (or there should not be any) fundamental constants.

As far as I can see, there are two lines we can follow to achieve this task. The first is to engage in a 'heavy-weight' metaphysical project with the aim to discover the relevant notion of fundamentality through the elucidation of what fundamental constants *really* are. To my knowledge, the core ideas behind the two main notions of fundamentality currently available are the following: First, the ontological independence (or metaphysical ungroundedness) of fundamental entities from the non-fundamental ones and, second, the characteristic feature of the former to constitute a minimal collection of entities which either individually or collectively provide a complete metaphysical ground for the latter. It is not at all clear whether the case of fundamental constants conforms to either of the above notions. A satisfactory explanation of this fact is that the above-mentioned notions of fundamentality have been born and developed in metaphysical contexts where the notions used (supervenience, ground, ontological dependence, etc.) as well as the arguments discussed do not concern *kinds* of entities such as the constants appearing in laws of nature (rather, they concern entities of the world such as particulars and [sparse, natural] properties). This acknowledgement, however, leads naturally to the need of an ontological account of fundamental constants capable of at least providing answers to the following questions: First, what is the ontological category of constants? Second, what is the nature of the metaphysical relation (if any) between laws of nature and physical constants?

The second line is to look into scientific practice to find a metaphysically 'light-weight' approach stemming from the way physicists themselves handle the issue and related to the constants' theoretical role. Here, I'll follow the second line and try to find some core elements that characterise the physicists' notion of fundamental constants and constitute its definition. My choice on this issue is not based on a prior belief about the redundancy of the aforementioned 'heavy-weight' metaphysical project. On the contrary, I think that the central role fundamental constants play in our scientific image of the world suggests that they are ontologically indispensable and so turns the articulation of a proper ontological account of them into a desirable metaphysical task to be accomplished. The choice of the 'light-weight' course has to do exclusively with the purposes of the present work. As we'll see in the sequel, we can argue for the efficiency of the argument from fundamental constants without delving into the hard issues related to their ontological status and their metaphysical relations with other entities. We may rest content with (a) the core notion of fundamentality of constants which physicists appeal to, and (b) the recognition of the uncontroversial fact that constants are 'ingredients' of laws, in the ontologically neutral sense that when the former change (due to the alteration of either their value or their functional form) the latter change as well (see also the remark in fn. 26).

To begin, here are some typical beliefs about the notion of the fundamentality of constants expressed by some prominent theoretical physicists:

> Fundamental constants are parameters that cannot be calculated on the basis of other constants, not just because the calculation is too complicated but because we do not know of anything more fundamental. (Weinberg et al. 1983)
>
> A fundamental constant is a parameter whose value we must supply in order to specify the Lagrangian of the Standard Model.... Fundamental constants are not reducible into more basic elements. (Wilczek 2007)
>
> Fundamental constants are parameters that cannot be calculated with our present knowledge of physics. (Uzan 2002)
>
> Fundamental constants are non-determined parameters which appear in the formulation of physical laws and we can only measure their value and not predict it. (Uzan & Leclercq 2008)
>
> Fundamental constants are fixed and universal values of some physical observables (dimensional or not) that enter the theories designed to describe the physical reality. (Cohen-Tannoutji 2009)

What I think is made clear from the above quotations is, first, that physicists themselves associate[27] the notion of the fundamentality of constants

[27] For a scientifically informed metaphysician, this association fits nicely her core belief that the only reliable source for discovering the inventory of fundamental entities of the world is our best current physical theories. Of course, there are disagreements about the appropriate theoretical frameworks, but what I think is undisputed is that the examination of the case of fundamental

with their role in a *particular* theoretical framework. This explains why they think that *which* parameters/observables are to be considered as fundamental constants depends on the choice of a framework comprising fundamental theories. The Standard Model of fundamental interactions (perhaps supplemented with Einstein's General Theory of Relativity) seems to be the most appropriate framework nowadays, but there are many physicists who choose other frameworks such as Grand Unified Theories or even String Theory. Furthermore, the crucial characteristic of fundamental constants (the one that according to physicists constitutes their definition) is the following:

> (FUNDCON) *Fundamental constants are parameters/observables which we cannot calculate on the basis of other constants or predict their value in a broader theoretical framework.*

We may discern a strong and a weak interpretation of FUNDCON. According to the former, it refers to both current *and* future theoretical frameworks, while according to the latter, it refers only to *current* theoretical frameworks. Under the weak interpretation, the definition leaves room for the existence of fundamental constants the values of which will be specified by a *future* theoretical framework. I think that the proper interpretation for our discussion is the strong one; for, it is doubtful whether the weak construal (which is based on a temporally relativised criterion of fundamentality) can support a metaphysically robust enough notion of fundamentality of constants (even within the metaphysically 'light-weight' context I have chosen to work).

Granted FUNDCON, let me first discuss how Bird himself defends the aforementioned objection to premise (1). He makes use of the definition in order to overcome the 'constants' problem by claiming that the constants appearing in laws of nature are *not* fundamental. To this end, he examines the case of the alleged fundamental constants that appear in non-fundamental laws (like the constant G in Newton's gravitational law). He points out that, despite appearances, these constants might

constants within a science-sensitive metaphysical context should be grounded in the findings and practice of *modern physical* science.

not be fundamental, since there is an epistemic possibility that within a broader, to be developed in the future, theoretical context their values will be constrained by fundamental laws. Yet, the justification of Bird's suggested possibility is certainly weak, as it is based on mere speculation about the future development of physics. Bird appeals to the scientific authority of Steven Weinberg (1992) in order to strengthen his view that there are no fundamental constants whatsoever. Weinberg (*ibid.*, Chaps. 9 and 10) believes that the history of modern physics indicates that there is a final Theory of Everything (TOE, for short), and that we are capable of discovering it. In the context of this theory, he predicts that all physical fundamental constants (with the possible exception of the cosmological constant) shall be *determined* by some kind of symmetry principles in such a way that any small variation of the value of any fundamental constant would destroy the consistency of the theory. Weinberg's belief, however, is based on mere speculation about the future of physics too.

Are there any reasons based on *current* science to insist that there are no fundamental constants? There is prima facie one such reason related to the ongoing debate about the number of fundamental *dimensional* constants necessary for the description of physical reality. For instance, Lev Okun (in Duff et al. 2002) argues that the fundamental dimensional constants are three, namely the velocity of light c, Planck's constant h and the gravitational constant G. While Gabriele Veneziano (*ibid.*) insists that, in the context of String Theory, there are only two fundamental dimensional constants, c and λ_s (string length). Finally, and prima facie crucially for the justification of Bird's response, there is a view defended by Michael Duff (*ibid.*; 2004), according to which the number of 'fundamental' dimensional constants is zero. On this account, only *dimensionless* constants can be literally regarded as fundamental, while dimensional constants are human constructs as they are simply conversion factors from one system of units to another. People who defend the existence of fundamental dimensional constants lay emphasis on the fact that the latter are fundamental because they define the limits of application of new fundamental theories and the appearance of new phenomena (Duff et al. 2002). They also exploit some recent experimental data showing, arguably, that one dimensionless constant, the fine structure constant α, varies with time. If this is true, they claim, it makes sense to ask which dimensional constant from among those entering its definition is responsible for the variation. And this seems to support the belief that dimensional constants cannot be human constructs. Of course, in order to ground this belief, one must show that the answer to the question 'which dimensional constant is responsible for the variation of α?' makes a difference to physics (Moffat 2002). This is, however, what Duff denies; he argues that by using the appropriate units (which can be determined experimentally without a reference to any constants) we can ascribe the variation of α to whichever dimensional constant we choose (of course, from among the ones that enter its definition) without any consequence for physics.

Where do all these leave us? One may think that Duff's view supports a direct vindication of Bird's objection to premise (1) of the argument. There are no fundamental dimensional constants and, consequently, it is not true that some laws of nature involve them. Yet, the difficulty strikes back, this time in the form of the *dimensionless* constants which, even for Duff, *are* fundamental and appear in the fundamental laws of nature according to our best theories. Currently, there are 31 fundamental dimensionless physical constants required by particle physics and cosmology. Twenty-seven of them appear explicitly in the relevant Lagrangians, whereas the remaining cosmological ones are inserted as initial conditions for our universe (for details, see Tegmark et al. 2006). So, if we think that Duff is right, the most we can get—and consistent with the current (and most probably, if we agree with Duff that physics concerns dimensionless quantities, the future) theories of physics—is the rejection of the view that there are fundamental dimensional constants.[28] This leaves us, however, with the problem of the fundamental *dimensionless*

[28] We may arrive to the same conclusion following Levy-Leblond's interesting views about the fate of all dimensional constants. Levy-Leblond (1979) classifies physical constants into three distinct types: Type-A includes properties of particular (types of) physical objects (such as the masses of elementary particles); type-B constants characterise classes of phenomena (such as the coupling constants of fundamental interactions); and type-C constants characterise the most general theoretical frameworks in the context of which we can describe any physical phenomena (such as ħ). According to Levy-Leblond, all type-C constants undergo a shift (as theories evolve) from a dominant conceptual status to an almost invisible one. They are 'progressively incorporated into the implicit common background of physical ideas, then play a role of mere unit conversion factors and often finally forgotten altogether by a suitable redefinition of physical units' (1979, 246).

ones. If, on the other hand, we think that Duff is not right, there are fundamental dimensional constants (and fundamental dimensionless ones as well). Hence, in any case, the truth of premise (1) is not compromised. I conclude that we do not have any cogent reasons to think that there are no fundamental constants.

7.3.2.2 Objection to Premise (2)

One may object to premise (2) by claiming that:

> *There is no metaphysically possible world in which the values of fundamental constants are even slightly different.*

How can the advocate of DEAL plausibly infer (via arguments based on, or at least inspired by, recent scientific findings) the metaphysical necessity of the values of fundamental constants? Here is a suggestion: She may defend the view that fundamental constants are metaphysically necessary by appealing to a special kind of arguments which, in their strong form, purport to explain the fine-tuning of the values of fundamental constants that characterise our observable universe. These arguments (which are most often called *anthropic*) also appear in a weak form, in defending the view that the values of fundamental constants constitute *necessary* conditions for the existence of observers. Weak anthropic arguments are supported by the Weak Anthropic Principle (WAP), according to which the necessary conditions for the existence of observers select out of all possible universes a subset which allows for the existence of observers. It is currently popular to make use of a theoretical hypothesis—the *multiverse* hypothesis[29]—to provide the required set of possible universes in

[29] The multiverse proposal is not a theory but a ground hypothesis (or prediction) of several modern explanatory physical theories in cosmology and high-energy physics. The core idea is that our universe is just one of a (possibly infinite) ensemble of 'parallel' universes; the latter could differ from ours in various features, ranging from different initial conditions (at their Big Bangs) to different fundamental laws and constants. In the cosmological field, there are various views on how the universes of the multiverse might arise. The most popular one nowadays is provided by *inflation* theory: under the assumptions of a spatially infinite universe and a uniform distribution of matter at large scales, a consequence of inflation theory is that our observable universe is part of a 'bubble' which underwent an extra-fast expansion phase at some early time. There are many other 'bubbles',

order for the WAP to be operative and explanatory. Though the multiverse hypothesis was not initially motivated as a possible ground for the anthropic explanation of the apparent fine-tuning of the values of constants, it is now presented as the key point which helps anthropic explanations to meet the objection that they are not scientific.

Clearly, the WAP cannot support the metaphysical necessity of the values of fundamental constants. There is a stronger principle, however, which prima facie purports to do exactly that. According to the Strong Anthropic Principle (SAP), the universe (hence the fundamental parameters on which it depends) *must* be such as to admit the creation of observers within it at some stage of its evolution. Once again, multiverse proposals may provide the relevant inventory of universes in order for the necessity following from the above "must" to be understood (without the aid of a Great Designer[30]). Granting the huge number of universes in the multiverse, how can we understand SAP's modal claim? To answer this question, we must first point out that there are two ways to comprehend the notion of multiverse. According to the first (inspired by the cosmological inflation theory; see fn. 29), the multiverse consists of a perhaps infinite number of *actual* universes, one of which is ours. While according to the second (inspired by the landscape version of String Theory), the multiverse consists of an infinite number of *possible* universes, one of which is the actual world. Starting with the latter interpretation, we note that in this case SAP's claim requires that *all* possible universes are characterised by fundamental constants taking values that allow for the creation of observers. But this is problematic; if, on the one hand, only

each with the same laws but different initial conditions which are most possibly created by quantum fluctuations during the period of inflation (Guth 1981). Another proposal on how new universes may arise is a variant of the above theory, the so-called *eternal inflation* scenario, according to which each 'bubble' universe is continually self-reproducing (Linde (1986), Vilenkin (1983)). For other cosmological theories based on varieties of the multiverse hypothesis, see Wheeler (1974), Misner, Thorne and Wheeler (1973), G.F.R. Ellis (1979) and Smolin (1997). Recently, the multiverse hypothesis has been used in high-energy physics as a ground assumption of the new version of String Theory, the so-called *Landscape Scenario*. For a recent volume dedicated to the multiverse hypothesis, see Carr (2007).

[30] Of course, if one is willing to explain necessity as discussed here with the assistance of a Great Designer, she may follow a theistic version of SAP, according to which there exists *one* possible universe especially designed with the goal of generating and sustaining (human) observers. In that case the claim for the metaphysical necessity of the values of constants is trivially true.

one set of values for the constants is the proper one, then SAP's claim itself undermines its explanatory power because it stands opposed to the core tenet of the multiverse scenario, according to which the constituent universes may be characterised by different values of constants. If, on the other hand, multiple sets of values are appropriate for observers to exist, then the *actual* set is not metaphysically necessary.

Let me now turn to the former conception of the multiverse. To understand SAP's claim in the context of that conception, we have to consider possible *multiverses,* each consisting of many universes. SAP then requires that, in each possible world, there is a multiverse such that it contains at least one universe with the right values for observers to exist. Clearly, this time, SAP is consistent with the multiverse hypothesis, because in each possible multiverse we may have different values of the constants in different constituent universes. It falls short, however, of providing grounds for the metaphysical necessity of the actual values of fundamental constants. Necessity can only be proved (via SAP) by assuming that the proper (for observers to exist) set of values is *unique;* yet, although a small change in a fundamental constant might so alter the universe that life could not evolve, compensatory changes could be made in the values of other constants to restore a set of favourable situations. Hence, it is most possibly not the case that there is only one set of values consistent with the existence of observers (see, for example, Stenger (2004), Harnik et al. (2006) and Adams (2008)).[31]

There is an alternative course that a scientifically informed metaphysician might follow to defend the view of the metaphysical necessity of fundamental constants. She might insist on the truth of the following claims which jointly could prove her view. The first two claims concern the existence of a unique TOE and its alleged ability to completely determine the

[31] The logic of anthropic reasoning does not indicate any special preference for human (or intelligent) life. Furthermore, there is disagreement about the scope of the notion of life and the elements which are capable of being the building blocks for the type of complex systems that might develop life-like features. Much of the argumentation in this area concerns cosmological and astrophysical features that could make life (broadly conceived) more likely. Typical habitable universes are characterised by big bang nucleosynthesis and large-scale structure, and allow for star formation, long stellar lifetimes and reasonable means to produce and disperse heavy chemical elements into the interstellar medium.

values of all fundamental constants, while the last one is a metaphysical assumption needed to ensure the metaphysical necessity of the upshot.

(i) There is only *one* TOE that describes the actual world.
(ii) TOE's fundamental laws and/or basic principles *uniquely* and *completely* specify the values of all 'fundamental' constants. There are no constants that can take different values without affecting the inner logic and self-consistency of the TOE.
(iii) If the value of a 'fundamental' constant is uniquely and completely specified by TOE's laws and principles, then it is metaphysically necessary.

Before examining the premises of this argument, it is important to note what I think is a serious drawback of it. Under the strong interpretation of FUNDCON, there is a tension between the latter and the claim, appearing in (ii) and (iii), that there can be *fundamental* physical constants the values of which can be predicted within the framework of the future TOE (that tension explains the inverted commas in (ii) and (iii)). In other words, if someone construes the definition of fundamentality as referring to both current *and* future theoretical frameworks, then (ii) and (iii) can refer to non-fundamental constants only. This is not an unwelcome consequence, *provided that the objection to premise (1) is true.* But if my previous arguments are sound, we have no cogent reason to rule out the existence of fundamental constants. Hence, given the strong interpretation, the most one can prove based on the assumptions (i)–(iii) is that *some* physical constants have metaphysically necessary values. But clearly that's not enough, because the argument against DEAL can work in that case by taking into account just the subset consisting of the *fundamental* constants.

The obvious escape route for the objector is to follow the weak interpretation of the definition of fundamentality. But, as I have already remarked, in that case she would ground her argument on metaphysically too thin a notion of the fundamentality of constants. Nevertheless, in order to grant the objector the best chances of countering the argument from fundamental constants given FUNDCON, I suggest to temporarily bracket my qualms and allow her (for the remainder of the discussion of the objection to premise (2) only) to articulate her argument in the context of the weak interpretation of the definition.

Let me begin, then, with the first claim; though a shared belief (or, better, a great hope) of most theoretical physicists, it is clearly speculation. Surely, it is a very fruitful one, guiding as it is the attempts of generations of physicists working on projects of unification of fundamental forces. But, nevertheless, there is no compelling evidence that there could be a TOE. Here, it might be objected that current physics provides a very promising candidate for a TOE, namely, String Theory. The majority of theoretical physicists working on fundamental issues have recently devoted all their efforts to the development of this theory, which was designed right from the start to achieve the unification of forces. Unfortunately, String Theory has its own problems and has been (inter alia) the target of extensive criticism for its inability to provide testable predictions within the range of even remote energy levels (see Smolin (2006) for a critical look). Not surprisingly, there is an ongoing debate not only about the appropriateness of String Theory as a candidate for a TOE, but even about its ability to meet the relevant standards in order to be classified as a *physical* scientific theory (and not just a mathematical one). In any case, at the current stage of scientific development, String Theory does not seem to constitute a cogent reason to believe in the existence of a TOE. And it certainly gives us no evidence that, were the latter to exist, it would be unique.

Based on the history of physical science so far, and recent developments in the only currently available candidate for a TOE (String Theory), I think that we are entitled to say that the second claim is also mere speculation. Despite physicists' hopes, it seems that the hitherto advancement of physical science has not resulted in a significant reduction of the number of fundamental constants. On the contrary, contemporary physical theories (such as the Standard Model of particle physics and modern cosmological theories) abound with constants which, though extremely significant, are not theoretically explained. As Lee Smolin points out, despite expectations,

> it seems that unification has simply not brought with it any understanding of how the parameters of the laws of physics were chosen. (1997, 70)

A large number of theoretical physicists, however, still believe that the hitherto unexplained values of all fundamental constants will be deducible from the basic principles of a (or, the) future TOE. Yet, recent developments in the unique candidate for a TOE so far not only fail to support this belief but suggest instead just the opposite. In particular, the most recent version of String Theory (the Landscape Scenario) strongly suggests the existence of a huge number of other possible universes corresponding to different solutions of its equations. The reason is that there is no unique way to compactify the (required for consistency) six extra dimensions of the ten-dimensional space of String Theory and thus produce a unique vacuum state describing the one and only universe. In the Landscape version of String Theory, different universes correspond to different minima of the vacuum energy and are characterised, inter alia, by different values of fundamental constants. Hence, claim (ii) fails because it makes no sense for a final TOE (conforming to the tenets of the Landscape Scenario) to *predict* the values of all fundamental constants since they can take different values in different universes. The best it can do is to identify a range of values of fundamental constants and/or determine the probability that a universe is characterised by constants taking specific values (or values falling within a specific interval). The obvious objection to the above argument is that the very hypothesis of the Landscape Scenario is a wild speculation concerning the foundations of the controversial String Theory. Yet, the Landscape hypothesis can earn a degree of credibility by being a specific form of the multiverse hypothesis which proved to be a fruitful assumption of several explanatory physical theories (see fn. 29); I conclude that, even if we grant that String Theory is a candidate for a TOE, we have no reason to assume (at the current stage of its development) that its fundamental laws and/or basic principles can *uniquely* and *completely* specify the values of all 'fundamental' constants.

Thus far I've remarked that, given the history of modern physical science and current scientific findings, it is not plausible to suppose that a future TOE (if any) would specify the values of all fundamental constants. But what if we suppose that, despite my arguments, a future TOE will do exactly that? If we grant the truth of (i) and (ii), can we infer the metaphysical necessity of the values of all constants? To answer this question, we must examine the last of the suggested claims. Bird introduces claim (iii) in order to justify the leap from the non-fundamentality of constants (according to the strong interpretation of FUNDCON) to the metaphysical necessity of their values. What reasons can we have to assume the truth of (iii)? Bird is not clear on this point, but I can guess (from what he says) two reasons that he might possibly provide.

Consider first the example of the constant G that appears in Newton's gravitational law. According to Bird (2007, 212):

> it might be, for all we know, that the value of G is necessary, despite appearances to the contrary, just as the fact that salt dissolves in water is necessary, despite initial appearances.

Here Bird draws on an alleged analogy between his well-known suggestion that some non-fundamental laws concerning the behaviour of substances are metaphysically necessary and the case of the constant G. But is this move justified? To answer this question, let me first review briefly the steps of Bird's original argument (2007, 177–8). Suppose that FL is a fundamental law and HLL a higher-level law which relates substances S_1, S_2 and expresses a relation R between them that supervenes on FL. Assuming that the constitution of S_1 and S_2 is metaphysically necessary, FL is necessarily sufficient for R to hold. Suppose further that the existence of at least one of the substances (e.g., S_1) depends on fundamental laws that include FL. Assuming again the necessity of constitution, the last assumption implies that, necessarily, FL is a necessary condition for S_1 to exist. Then, there is no possible world where the substance S_1 exists but is not related in the way HLL says it is (i.e., through R). Hence, HLL is metaphysically necessary in the sense that it holds in all possible worlds in which its relata exist. Given that the premises are independent of whether FL is metaphysically contingent or necessary, the argument purports to show that even a contingentist about fundamental laws must concede that *some* laws are metaphysically necessary.

Let me now illustrate how one might apply Bird's argument in the case of fundamental constants by examining the particular case of the gravitational constant G. In the original argument, the fundamental law FL is Coulomb's law and the higher-level law HLL is the law that salt dissolves in water. Now, in the case of G, FL must be a relevant fundamental law of the TOE and HLL may be Newton's law of gravitation expressing a relation R between the *specific actual value* of the constant G, the masses m_1 and m_2 of two bodies and the distance r between the latter. In formal terms, the analogous argument runs as follows (where SAVG is the proposition that the constant G exists and takes its specific actual value):

(a) $(\neg HLL \rightarrow \neg R)$
(b) $(\neg R \rightarrow \neg FL)$
(c) $(\neg HLL \rightarrow \neg FL)$
(d) $(\neg FL \rightarrow \neg SAVG)$
$\therefore$ $(\neg HLL \rightarrow \neg SAVG)$

The conclusion of the argument is that Newton's gravitational law is metaphysically necessary, in the sense that it holds in all possible worlds where its relata exist. But this does not prove that the actual value of G is metaphysically necessary. It merely shows that there is no possible world in which G takes its actual value and the gravitational law does not hold. Actually, one cannot even prove the conclusion of the argument, due to the lack of justification for premise (d). In the original argument, the premise is justified via the assumption (grounded in independent arguments for the necessity of constitution) that what holds in the actual world (namely, that FL is a necessary condition for S_1 to exist) holds in any other possible world (in which S_1 exists) as well. But this is not the case in the corresponding premise of the analogous argument. The fundamental laws of the TOE may uniquely predict the value of G but there is no justification for the claim that, necessarily, different fundamental laws would predict different values for G.

Nevertheless, one may take a cue from the other example that Bird presents (the law of the intensity of light) to provide a reason for the truth of his contentious assumption. Bird points out that we can imagine a world where the intensity of light is not proportional to d^{-2} (where d is the distance from the light source). But, for Bird, this does not mean that the value of the exponent is metaphysically contingent. For, the fact that the intensity is proportional to d^{-2} is derivable from another, more general law, the law of conservation of energy, and a world that does not respect that is not at all similar to the actual one because, within it, energy is not conserved. As I understand him, Bird objects to the metaphysical contingency of the exponent on the basis that energy is *essentially* conserved; hence, a world with a non-conserved energy-like characteristic (despite appearances) does not contain energy and so is not similar to ours. Returning now to the case of constants, we may argue similarly; given that the value of the constants may be derivable from deeper, fundamental laws, a world characterised by different values is not similar to the actual. For, that world cannot be populated by the actual properties, since the latter are essentially such that the fundamental laws hold. The upshot of this reasoning is in effect that (despite appearances) it is not metaphysically possible to have both the actual properties *and* different values for the constants. This, however, is a direct objection to premise (3). So, to evaluate the plausibility of the above reasoning, we have to take a closer look at the last premise of the argument.

7.3.2.3 Objection to Premise (3)

The main objection to the last premise of the argument is the following:

> *Even the slightest difference in the values of the actual fundamental constants requires that the properties appearing in the laws be different from the ones inhabiting the actual world.*

Within a DEAL-friendly context, this claim can be grounded only in a *fine-grained conception* (FGC) of the dispositional nature of properties, according to which each law with a slightly different constant involves a different property of a (slightly?) different dispositional nature. The suggested co-variation of properties and constants is of course entirely in the spirit of Dispositional Essentialism. Indeed, it is an application of the well-known dispositional essentialist 'different laws–different properties' strategy. If fundamental constants take different values, the laws involving them must change. But according to dispositional essentialists, we cannot have different laws unless we have different nomically related fundamental properties. From the perspective of FGC, the value of the constant must be involved in the dispositional essence of the property and, therefore, in its transworld identity. Hence, though the values of the constants are not unique, there can be no independently varying fundamental constants; for, any variation would mean that we no longer have the same properties.

Can the proponents of DEAL plausibly claim that FGC constitutes the ground of a serious objection to premise (3)? In what follows, I'll present my worries about thinking of FGC as the ground for the 'different laws–different properties' strategy. Friends of DEAL are challenged to meet these qualms; otherwise, they cannot justifiably ground their strategy against premise (3) of the argument from fundamental constants.

Let me begin by presenting a case which, though it does not create insurmountable difficulties for FGC, raises, nevertheless, issues that obviously need clarification. Furthermore, I think that the proper accommodation of this case within the context of FGC increases significantly the complexity of the latter rendering it less persuasive. Let us then consider the case of a law involving several dispositional properties/relations. This is a plausible case especially from the perspective of Dispositional Monism. Since dispositional monists insist that *all* properties and relations are dispositional, and since laws are relations between properties/relations, there can be no law with only one dispositional property in the monistic context. (Most often one of the relata of the nomic relations is a spatiotemporal relation which, according to Dispositional Monism, is also dispositional.) The problem for the proper implementation of FGC is to specify *which* property of the nomically related ones involves in its individual essence the value of the fundamental constant. I suggest two alternatives on behalf of the proponent of FGC: First, she may check other laws that involve the same constant and see which of the properties in question (if any) appear in these laws. This strategy, however, may not succeed, because the constant may appear in only one law involving the relevant property.[32] A way out for the proponent of FGC is to argue that the fundamental constant is involved in the individual essences of *all* properties that appear in the law (this claim, for instance, can be justified

[32] Note here that, though a specific constant may be involved in the individual essence of the property, the latter may also appear in laws without constants or with different constants.

within a context where it is assumed that the law expresses the essence of properties which are reciprocal disposition partners[33]). This suggestion, however, solves the problem by increasing the complexity of FGC. The complexity may be further increased in the case of a constant appearing in several laws, because in that case it must be involved in the individual essences of all the other (nomically related) properties as well. Examples of such constants are the fundamental dimensional type-C constants (such as c and ħ) which, according to FGC (as we now construe it), must be involved in the individual essences of a number of properties appearing in a variety of different laws in different theoretical frameworks.[34]

A second difficulty for FGC emerges from the existence of constants with values depending on the number of spatial dimensions of the world. It can be shown that the values of some *dimensionless* constants (constructed out of dimensional ones) depend on spatial dimensionality. For instance, in d spatial dimensions, the dimensionless constant constructed from G, h, c and e is proportional to $h^{2-d} \cdot e^{d-1} \cdot G^{(3-d)/2} \cdot c^{d-4}$ (Barrow and Tipler 1986, 269). Another example is the case of numerical constants of geometrical origin that appear in various laws. These constants also take different values in universes with different spatial dimensionality. Now it seems that if the dimensionality-sensitive constants change, then the laws in which they appear should change as well. So, we can have the same properties appearing in different laws with different values of constants.

To illustrate the point, consider, for instance, the case of the numerical constant appearing in the law $L_@$ determining the electric field E(r) of a point charge q in three spatial dimensions:

$$E(r) = \frac{1}{4\pi}\frac{q}{r^2} \quad (L_@)$$

[33] For the notion of reciprocal disposition partners, see Martin (2008).

[34] Recall (from fn. 28) that type-C constants characterise the most general theoretical frameworks in which we can describe physical phenomena. Since each of these constants is present in most of the laws of the framework that it characterises, it should be involved in the individual essence of a large number of fundamental properties. Furthermore, since fundamental properties appear in laws of different theoretical frameworks, they should involve in their individual essences all the type-C constants that characterise the frameworks they appear in.

Starting from Gauss' law $\nabla \cdot E = \rho$, which is valid in all spatial dimensions (provided, of course, that we use the appropriate definition of the gradient), we can prove that the value of the electric field for a point charge in a world with d spatial dimensions is given by

$$E(r) = \frac{\Gamma\left(\frac{d}{2}\right)}{2\pi^{\frac{d}{2}}} \frac{q}{r^{d-1}} \quad (\mathrm{L}_*)$$

where Γ(x) is the gamma function and r the spatial distance from the point charge (Zwiebach 2009, 57). Consider now a possible world $w_\#$ with spatial dimension $d_\# \uparrow 3$. Application of the above general formula in this case yields

$$E(r) = \frac{\Gamma\left(\frac{d_\#}{2}\right)}{2\pi^{\frac{d_\#}{2}}} \frac{q}{r^{d_\#-1}} \quad (\mathrm{L}_\#)$$

In $w_\#$, $L_\#$ is a law that differs from the actual $L_@$; so, it seems that there can exist two different laws with different values of a constant both involving the electric charge q and the spatial distance r.

The proponent of FGC might most possibly wish to address this difficulty for DEAL by applying the different laws–different properties strategy. But, as I shall now argue, the three most reasonable ways to implement the strategy fail to respond to my challenge from the dimensionality-sensitive constants.

First Response $L_\#$ characterises $w_\#$ and is different from $L_@$. This fact, however, is compatible with the FGC because $L_\#$ does not relate q and r. Here the advocate of DEAL, depending on her solution to the previously mentioned 'which' problem, may insist that $L_\#$ involves $q^\#$ instead of q, or $r^\#$ instead of r, or both $q^\#$ and $r^\#$ instead of q and r, respectively.

The tenability of this first response depends on what grounds the advocate of FGC has to insist that, in $w_{\#}$, the nomically related properties/relations are not q and r. A popular strategy is to appeal to modal intuitions. DEAL proponents have repeatedly argued that it is difficult to *conceive* how the same property might have a nomic/causal role different from the one it actually has. (The argument most often highlights the inconceivability of a swapping-of-roles scenario; for instance, it is difficult to conceive how the electric charge might have the actual causal role of mass and vice versa.) The strength of modal intuitions, however, depends on the *degree* of difference between the actual and the possible role. Perhaps one may grant dispositional essentialists that it is difficult to conceive two identically charged bodies attracting each other. But is it equally difficult to conceive of the property of charge as obeying a law of electrostatic force with a *slightly* different (from the actual) value of the constant[35] appearing in the law? I think it is not; modal intuitions are not strong enough here. So the appeal to them is not a powerful strategy in this case.

Do proponents of FGC have any a posteriori grounds (suggested by findings of modern science) to insist on their response? According to my scenario, $w_{\#}$ differs from the actual world *only* with respect to the number of spatial dimensions. Hence, they must give us a cogent reason to think of this scenario as metaphysically *impossible*. In line with their view, this alleged impossibility should be based on the claim that the individual essence of fundamental properties/relations involves somehow the number of spatial dimensions of the world they 'live' in. This *could* be a tenable metaphysical claim as far as the spatial distance relation is concerned. One might argue, for instance, that the nature of the latter, *qua* spatial distance, is metaphysically dependent on the number of spatial dimensions. But clearly the advocate of FGC cannot rest content with that. Some dispositional essentialists who may embrace FGC are dualists and think that spatiotemporal relations do not have any dispositional essence. Furthermore, even if we assume that all advocates of FGC are dispositional monists, the thesis that FGC holds only for *one* kind of

[35] Variation of the constant results in this case from its functional dependence on d which, in turn, is due to the geometric origin of the constants under consideration.

fundamental features of the world seems quite implausible. Hence, the proponent of FGC should provide a reason for suggesting, at least, that the individual essence of other fundamental properties/relations (besides spatial ones) involves somehow the number of spatial dimensions. Here I think lies the difficulty; for I cannot see how she can do that.

A dispositional essentialist may think that, despite my arguments to the contrary, she can follow a science-informed line of thought which arguably leads to the conclusion that there is an implicit metaphysical relation between the number of spatial dimensions and fundamental properties. The impression that there is such a relationship can be due to the observation that String Theory *determines* the number of spatial dimensions of the world, in the sense that its very consistency requires that the number of spatial dimensions of the universe is nine. The relevant argumentation may be fleshed out as follows. Recall first that, according to QFT, the strength of each fundamental force depends in part on the actual value of the associated coupling constant. Now, according to the dispositional essentialist account, the identity of each fundamental property is exclusively determined by its actual causal roles. But surely, the causal role of a fundamental dispositional property involves somehow the strength of the relevant fundamental force which is its manifestation. Hence, given the dependence of a force's strength on the corresponding coupling constant, it seems that the dispositional account implies that the identity of fundamental properties must depend somehow on the values of some fundamental constants.[36] Given the intimate relationship between identity and essence, dispositional essentialists may then suggest that the values of coupling constants *must* be involved somehow in the essence of the relevant fundamental properties.[37] Note however that, according to String Theory, 'fundamental' coupling constants are determined by other constants which reflect the geometry and topology of the space corresponding to the extra spatial compactified dimensions. Hence, given that

[36] Of course, we may arrive at the same conclusion by considering the role of fundamental constants in non-fundamental laws. For instance, according to Newton's gravitational law, the strength of the gravitational force depends on the value of the fundamental constant G. So, according to dispositional essentialists, the identity of mass must somehow depend on the value of G.

[37] Despite its initial appeal, this claim is not defensible after all. To see why, recall my argument from the application of the renormalisation procedures in QFT.

coupling constants are prima facie involved in the individual essences of fundamental properties, a dispositional essentialist might argue that we have an *implicit* metaphysical relation between nomically related fundamental properties and the topology/geometry of the space formed by the compactified dimensions.[38]

Even if we grant, for the sake of the argument, that String Theory (despite the problems I have already discussed) is the best currently available candidate for (or provides the paradigm of) a future TOE and so is (for the scientifically informed metaphysician) the most reliable source for drawing metaphysical conclusions, the above conclusion falls short of proving what the proponent of the FGC wants to prove. Recall that the desideratum is to show that the individual essence of fundamental properties involves somehow the number of spatial dimensions of the universe. But all the upshot of the preceding line of thought indicates is that the individual essence of fundamental properties involves somehow the kind of topology and geometry characterising the possible spaces formed by the extra spatial compactified dimensions of the world. Hence, despite first impressions, String Theory does not suggest that the global topological property of spatial dimensionality determines, or is determined by, the kinds of fundamental properties of the world. In the absence of an indication of such a kind of determination, even the scientifically informed metaphysician who thinks that String Theory is a paradigm of a TOE is not entitled to claim that the individual essence of fundamental properties involves somehow the number of spatial dimensions of the universe.

---

[38] In order to be persuasive, the reasoning should have the form of the following argument:

a) The individual essence of fundamental properties involves the values of coupling constants.
b) The individual essence of coupling constants involves the topology/geometry of the space with compactified dimensions.

∴ The individual essence of fundamental properties involves the topology/geometry of the space with compactified dimensions.
It is not clear, however, why the determination of coupling constants by 'topological/geometric' ones implies that the individual essence of the former involves the latter. Furthermore, the conclusion is only valid provided that a controversial principle of transitivity for individual essences holds.

Second Response Despite appearances, the actual world and $w_{\#}$ do not have *different* laws relating q and r. Rather, both worlds are characterised by one and the same *functional* law L_* relating q and r.[39] The spatial dimension d is the relevant argument of these functions and determines the specific form of L_* in each possible world.

A crucial metaphysical assumption of this response is that fundamental properties/relations have *functional* individual dispositional *essences.* It is not clear what having such an essence could mean. If, on the one hand, having a functional individual essence is tantamount to having a (perhaps infinite) conjunction of conditional individual essences (of the kind 'If the spatial dimension is d, then L_d is a law expressing a part of a property's individual essence', where L_d is the specific form of L_* in a world of spatial dimension d), then the problems of the preceding response arise again. Because now the advocate of FGC must find reasons to ground her claim that: (a) in the actual world, it is a part of the individual essence of q (or r, or both) the fact that it involves the value d=3; and (b) in $w_{\#}$, it is (another) part of the individual essence of q (or r, or both) the fact that it involves the value $d=d_{\#}$, and …., ad infinitum. If, on the other hand, having a functional individual essence means something *more* than the aforementioned conjunction, the burden is on the proponent of FGC to give a convincing account of the *ontological* correlate of this extra meaning.

Third Response In each possible world there exist all possible fundamental properties (including, of course, the actual ones) and the laws expressing their essence. Some of these properties and their associated laws are instantiated while others are not. The only difference between worlds concerns *which* properties and laws are instantiated. But, according to this view, this difference is not metaphysically relevant to the account of laws under consideration. What matters is that all possible properties are metaphysically necessary beings, and that in each possible world there exist (metaphysically necessary) laws expressing their essence. This seems to be the view that Bird (2012, 40) puts forward, and it works in our

[39] Laws such as $L_@$ and $L_{\#}$ are functional because the nomically related properties/relations q and r can take infinite values. (See, for instance, Armstrong's (1997, 242) account of such laws as relations between determinables having infinite—probably mostly uninstantiated—determinate values falling under them.) L_* is a functional law as well, though it has the extra argument d.

example in the following way. The fundamental properties/relations q and r 'corresponding' to the actual value of the relevant fundamental constant are instantiated in the actual world. Yet, there also exist a (perhaps infinite) number of *uninstantiated* fundamental properties 'corresponding' to all other possible values of the relevant constant. Something akin to this is true in $w_{\#}$; the only difference between the two worlds is that in $w_{\#}$ the properties instantiated are $q^{\#}$, $r^{\#}$ or both, instead of q, r or both, respectively. According to this proposal, there is no difficulty related to the presence of dimensionality-sensitive fundamental constants. In both worlds we have the *same* properties (some instantiated, others not) and the *same* laws (some instantiated, others not) expressing their essence, a result in complete agreement with the basic tenet of DEAL.

I am not sympathetic to this Platonic proposal, because I find the *actual* existence of uninstantiated properties and laws rather obscure. In any case, this is of no use to the critic of the argument from the existence of fundamental constants. The reason is that it begs the question against premise (2) of the argument by entailing the metaphysical necessity of the values of all actual fundamental constants. Given that all actual properties (instantiated and uninstantiated alike) exist in all possible worlds, the corresponding constants exist in all worlds as well, and so are metaphysically necessary.[40]

My final objection to FGC concerns again coupling constants and is related to a point which I have already stressed. Recall from Sect. 3.1 that the renormalisability of QFTs requires the functional dependence of the actual coupling constants on the energy level. Consider, for instance, the case of the fine structure constant α. Due to the phenomenon of vacuum polarisation (namely, the screening of an electron by a collection of electron–positron pairs produced by virtual photons that the electron emits), there is an increase of the electromagnetic coupling at high energies and the resultant constant is the effective or running coupling constant.[41] The

[40] The mode of expression presupposes the common assumption that all natural properties are transworld entities. Anyone who believes instead that they are world-bound must re-express the syllogism in terms of counterpart theory, where the *de re* modal representation is achieved by counterparts of properties, rather the properties themselves. Similar remarks hold for the use of the expressions 'same properties' and 'transworld identity'.

[41] The strong coupling constant α_s is also energy-dependent but, unlike α, decreases at high energies (asymptotic freedom). The difference from the electromagnetic case is due to the nature of gluons

formula which gives (in the case of the scattering of two electrons) the value of the running constant as a function of the momentum level Q is approximately

$$\alpha\left(Q^2\right)=\frac{\alpha}{1-\left(\dfrac{\alpha}{3\pi}\right)\ln\left(\dfrac{Q^2}{m_e^2}+1\right)}$$

where m_e is the electron's mass and $\alpha=\alpha\left(Q^2=0\right)$.

According to FGC, in any possible world where fundamental constants have slightly different values, there exist different fundamental properties. But if the advocate of FGC assumes that this is true transworldly, I cannot see what can prevent her to assume that it is true interworldly as well. Consider now a law in which the fine structure constant α appears. At each energy level of theoretical description, the coupling constant takes different values. Now it seems that if the constant changes, then the law in which it appears should change as well. So, we can have the same properties appearing in different laws characterised by different values of the constant.

Once again, the proponent of FGC would tend to meet this difficulty for DEAL by applying her favourite strategy. Just like her response to the second difficulty, she might implement the strategy in three different ways, all of which (as I shall argue) fail.

First Response Consider two energy (−momentum) levels, Q_1 and Q_2. Law L_{Q1} is different from law L_{Q2}. This fact, however, is entirely compatible with FGC because L_{Q2} and L_{Q1} do not relate the *same* fundamental properties.

Just like the first response to the second difficulty, the problem with this one concerns the grounds the advocate of FGC has to insist that at each energy level the nomically related properties/relations are different. First, the appeal to modal intuitions is quite controversial here. Can't we easily conceive of the *same* property as appearing in laws correspond-

(mediators of the strong force) which, unlike photons, are self-interacting.

ing to different energy levels? Indeed, this is what all physicists routinely assume when describing the energy-dependence of the strength of a force corresponding to a specific property. Fundamental properties such as electric charge, weak isospin, weak hypercharge and colour are related to fundamental interactions the strengths of which are energy-dependent due to the energy-variation of relevant dimensionless constants (α, α_w and α_s for the electromagnetic, weak and strong interactions, respectively). Nowhere is it suggested, however, that variation of energy level 'affects' the kind-identity of fundamental properties. So, not only can we conceive of the same properties as appearing in laws with different values of constants at different energies; in fact, as scientific practice suggests, it is difficult to find a posteriori grounds (suggested by findings of modern science) for the basic FGC claim in this case. The advocates of FGC must give us a persuasive reason to think that the individual essence of fundamental properties/relations involves somehow the energy level of the theoretical description. I cannot see, however, how they can support their claim on scientific grounds.

The direct implementation of the different laws–different properties strategy within the context of the first response raises a further problem. It compels the advocate of FGC to admit the existence of an extremely huge number of *actual* fundamental properties. It is important to note at this point that I do not think that this ontological inflation of fundamental properties raises a serious problem for any kind of multiverse scenario in which there *actually* exist a large number of universes[42]; perhaps, in such a case, an ontological inflation of fundamental properties is a *natural* consequence of the numerical inflation of the actual universes that constitute the multiverse (in each universe with different constants we have different properties). I strongly believe, however, that such an ontological inflation is very difficult to swallow when it concerns *our* observable universe (which perhaps is one of the actual constituents of a multiverse).

Second Response Despite appearances, all energy levels are characterised by one and the same *functional* law L_Q involving the same *functional* constant $\alpha(Q)$. The energy (−momentum) level Q is the relevant argument

[42] This is the case according to the cosmological inflation theory.

of these functions and determines the specific form of L_Q at each energy (−momentum) level.

The same problem that besets the second response to the second difficulty arises *mutatis mutandis* in the present case. The response presupposes that fundamental properties/relations have functional individual essences. According to an 'extensional' interpretation of such a functional essence as a conjunction of conditional individual essences (of the kind 'If the energy level is Q_i, then L_{Qi} is a law expressing a part of property's individual essence', where L_{Qi} is the specific form of L_Q at the energy level Q_i), the first two problems of the previous response arise again. Because in that case the FGC advocate must find reasons to ground her claim that, at *each* energy level Q_i, a distinct part of the essence of a nomically related fundamental property (or properties) is the fact that the latter involves somehow the specific energy value Q_i. If, on the other hand, the FGC proponent chooses to ascribe a 'non-extensional' meaning to the functional essence, the burden is on her to provide a convincing account of the *ontological* correlate of this meaning.

Third Response The last response I want to examine mimics (like the first two) the corresponding third response to the second difficulty. Fundamental nomically related properties/relations which correspond to specific values $a(Q_i)$ at each energy level Q_i are actually instantiated. Yet, there also exist a (perhaps infinite) number of *uninstantiated* fundamental properties 'corresponding' to all other possible values of the constant. According to this proposal, the problem with the energy-dependent constants does not arise. At each energy level we have the *same* properties (some instantiated, others not) and the *same* laws (some instantiated, others not) expressing their essence. The problems are familiar from the above discussion and, so, I will be very brief. First, we have the ontological inflation of actual fundamental properties we have already met in the first response. Second, we have the metaphysically obscure notion of uninstantiated fundamental properties. I think that the combination of these difficulties renders the third response rather implausible.

The totality of the difficulties presented above, taken jointly, undermine the attempt to ground the objection to premise (3) on the assumption of the fine-grained dispositional essence of properties. They also show that it is not plausible to defend (on dispositional essentialist grounds) Bird's claim (iii), according to which the unique and complete specification of the value of a 'fundamental' constant by TOE's laws and principles implies its metaphysical necessity. (Recall that the most reasonable defence of this claim is intimately related to the objection to premise (3).) Hence, one cannot object to premise (2) even if we grant that there is only one TOE which uniquely and completely specifies the values of all 'fundamental' constants.

To summarise our results: The advocate of DEAL may try to undermine the argument from the existence of fundamental constants by raising objections to all three of its premises. But I have argued that, contrary to her claims, we do not have any good reasons to believe that there are no fundamental constants. Furthermore, even if we grant the complete determination of the values of all fundamental constants by a future TOE, DEAL-proponents cannot plausibly argue that the actual values are metaphysically necessary. And finally, a fine-grained conception of the dispositional essence of properties cannot plausibly be the ground for the application of the different laws–different properties strategy in the case of fundamental constants. Given all that, the objections fail and the argument from the existence of fundamental constants remains a 'constant' threat to DEAL.

An important corollary of the above conclusion is that even dispositional monists have cogent reasons to hold that *some* laws of nature do not have an inferior ontological status. Recall that, according to them and their favourite account of laws (i.e., DEAL), all nomic relations are of the second type and are internal. It seems, however, that a qualification is needed at this point. For, there is nothing in the nature of these nomic relations which can preclude the presence of a fundamental constant as a relevant factor. And, as the present arguments indicate, nomic relations involving fundamental constants cannot supervene on their properties-relata, even if we grant that the latter are dispositional. (Note that in our view the only entities that can be relata of nomic relations are properties and relations. Taking fundamental constants as relata of nomic relations would render the supervenience of the latter a trivial fact.) In other words, I have concluded that there exist two *sub-types* of the type of nomic relations relating only dispositional relata: The first contains all nomic relations which do not involve any constants (or, if they do, the constants are not fundamental), and the second contains all nomic relations which do involve fundamental constants. A further upshot then is that even in a D-world, all members of the second sub-type, *qua* non-internal relations, can be genuinely existing entities.

8

Metaphysical Features of Nomic Relations and Laws

Are laws metaphysically contingent or necessary? Are nomic relations internal or external? These questions (and others related to them) constitute the core of an ongoing debate among metaphysicians. This chapter aims to contribute to this discussion.

8.1 The Modal Status of Laws of Nature

The kind of approach one endorses about the modal status of laws depends on (but, as we shall see, is not determined by) the preferred metaphysical account of (fundamental) properties. I'll start the discussion by briefly examining the current (standard and non-standard) views of categorical and dispositional monists. Traditionally, they have tended to endorse opposite positions regarding the modal status of laws. Though this is still the most popular tendency, there are dissenting voices showing that there is room for alternative approaches.

V. Livanios, *Science in Metaphysics*,
DOI 10.1007/978-3-319-41291-7_8

8.1.1 Categorical Monistic Approaches

Categorical monists (in their majority), often influenced by strong intuitions concerning the existence of possible worlds governed by different (from those in the actual one) laws, hold that laws of nature are metaphysically *contingent*. Consider, for instance, Armstrong's account of laws, according to which the relation of nomic necessitation holding between natural properties (universals in Armstrong's view) entails the corresponding regularity, yet is metaphysically contingent itself. The same fundamental natural properties may stand in a nomic relation in one possible world and fail to stand in that relation in another possible world.

The metaphysical contingency of laws in the categorical monistic context is supposed to 'flow' from the very 'nature' of categorical properties themselves. For it seems that there is nothing 'in' the latter which might dictate that specific properties must be (nomically) related to other specific properties. This fact, however, does not exclude the possibility of metaphysically *necessary* laws expressing external nomic relations between categorical properties. Evan Fales (1993, 140) is a well-known philosopher who tentatively suggests such a view. He acknowledges, however, that this account treats the metaphysical necessity of laws as primitive and sui generis. I agree; for, if nomic relations, *qua* external, do not supervene on the nature of their relata, what determines that it is the connections expressed by those relations that exist and not others? Nomic necessitarianism seems to be consistent with Categorical Monism, but, due to its primitive character and lack of independent motivation, it is far less believable than the orthodox (in the categoricalist camp) nomic contingentism.

In opposition to the above view, one may argue that there is an independently motivated ontological account that prima facie compels the categorical monist to embrace nomic necessitarianism. Armstrong (2004a), (following and slightly modifying Donald Baxter's (2001) work), puts forward the view that we may understand the controversial 'relation' of instantiation as a case of *partial identity* between universals and particulars. In Armstrong's words,

> Particulars are ones running through many different universals, universals are ones running through many different particulars. (2004a, 141)

This novel approach addresses the problems related to the proper account of instantiation construed as a genuine relation without invoking 'brute', ontologically fundamental, states of affairs, as Armstrong's (1997) earlier theory does. According to the instantiation-as-partial-identity account, a state of affairs <a is F> is a non-mereological partial identity between the particular a and the universal F. Particulars and universals are distinct but inseparable because they each help to *constitute* the other (*ibid*., 143). According to Armstrong (but not Baxter), a crucial consequence of the suggested account is that, given the actual (and metaphysically contingent) existence of specific universals and particulars, each actual state of affairs is *necessary*. Or, in other words, each actual instantiation of a universal is necessary and no other instantiation of it is possible (*ibid*., 148).[1] This account is supposed to hold for both monadic and relation-universals and, if we further assume that higher-order relations do not differ (in the relevant for the account modal aspects) from the first-order ones, we may claim that it also holds for the relations between the universals themselves. This last claim provides the link between the instantiation-as-partial-identity account and nomic necessitarianism. Nomic relations are relations between natural properties (which, in Armstrong's view, should be construed as universals); hence, according to the instantiation-as-partial-identity account (and given the above assumptions), they should be partially identical with the class of their relata, and (according to Armstrong's interpretation) the higher-order state of affairs resulting from the nomic relations relating their relata should be metaphysically necessary (see also Armstrong 2004, 136). Even if we grant that a nomic relation and the corresponding nomically involved properties/relations are contingent beings, if they both exist in a possible world then the state of affairs they constitute, *which is the relevant law,* also exists in that world.

Here I should make clear that the above conclusion follows immediately from Armstrong's account only if we embrace the transworld

[1] The instantiation-as-partial-identity account does not contradict Hume's dictum because we do not have necessary connections between *wholly* distinct existences.

identity account of the *de re* modal representation of properties and relations. Given that the very *same* nomic relations and their relata exist in various possible worlds, if they are partially identical in one world, they should (assuming the necessity of partial identity) be partially identical in the other worlds as well. If, however, one embraces the counterpart theory of the *de re* modal representation of properties and relations (as I do), she may prima facie accept both Armstrong's new account and (*pace* him) the view that each instantiation of a universal is metaphysically contingent. The instantiation-as-partial-identity account allows for this possibility, provided that the latter concerns *counterparts* of the particular and the universal. Extending this claim to higher ontological levels, we are prima facie entitled to say that the higher-order state of affairs constituted by a nomic relation holding between its properties-relata is metaphysically contingent (in the sense that there exist a counterpart of the nomic relation and counterparts of its relata such that the former is not partially identical with the latter[2]). But as argued in Livanios (2012a), nomic relations have essentially another feature which excludes the aforementioned possibility; nomic relations are (by their own nature) *relata-specific*. They do not only *relate* their relata; they also relate *their* relata. To illustrate, consider, for instance, a law of nature F(P, Q), where P and Q are fundamental natural properties or relations and F a nomic relation that actually relates them. Assuming that F is relata-specific, it relates P and Q as soon as it exists. Or, in other words, given that F holds between P and Q in the actual world, it holds between P and Q in all possible worlds in which F exists (F could not have existed and failed to relate P and Q). Co-existence with its relata is essential to F and this fact ontologically constitutes the unity of F(P, Q). Hence, though the instantiation-as-partial-identity account does not entail the metaphysical necessity of *all kinds* of states of affairs, it does entail it in the special case of laws (construed as higher-order states of affairs).

[2] Armstrong (2004a, 145, fn.7) discusses Hawthorne's worry about the grounds of the relevant counterpart relation. Hawthorne points out that Armstrong's theory compels the counterpart theorist to admit that, besides particulars, properties and relations should have worldbound counterparts in other possible worlds. Given that no two worlds share any particulars or universals, he wonders what would make a property in another possible world a counterpart of an actual property. I offered my answer in Sect. 4.3.4; that is, the counterpart relation is primitive.

Of course, all of the above remarks hold on the proviso that we buy the whole metaphysical package that Armstrong offers. That is, fundamental properties construed as universals and nomic higher-order universals which *instantiate* their lower-order properties-relata. Hence, at the very least, Armstrong's account does not threaten nomic contingentism in the categorical monistic context *in general*. One may further object to the assumptions that the above conclusion (stating the alleged metaphysical necessity of nomic states of affairs given the instantiation-as-partial-identity account) presupposes. For instance, no reason is given why we have to assume that the instantiation-as-partial-identity account holds for *higher-order* relations. Especially, in the case of higher-order *nomic* relations, it seems that we have no clear motivation to appeal to such a theory. I suggest that one main motivation for the theory is to provide a convincing account of the unity that characterises facts or states of affairs. Yet, in the case of those states of affairs which are laws this task can be achieved by appealing to the relata-specificity of nomic relations. Furthermore, it is not clear whether one can consistently apply the theory in the case of relation-universals in the first place. I find the partial-identity between relation-universals and *classes* of particulars obscure.[3] Finally, I have also qualms about the theoretical adequacy of the instantiation-as-partial-identity account, as the several difficulties that Armstrong himself tries to meet clearly indicate. For instance, I think that Armstrong's (2004a, 146–7) response to the charge that his theory blurs the distinction between universals and particulars cannot also be applied in the case of nomic relations. Armstrong presents the variable adicity and the non-repeatability of particulars (in opposition to the Instantial Invariance and the multiple instantiability of universals) as two features that might serve to distinguish universals from particulars in spite of the symmetrical treatment of them by his partial-identity account. Both features, however, fail to distinguish between nomic relations and their property-relata, since both are universals characterised by fixed adicity and capability of multiple instantiation. For all the above reasons, I conclude that Armstrong's

[3] Armstrong (2006, 244) forsakes this approach in favour of another account which first identifies the instantiation of a relation-universal by a class of particulars with the instantiation of the corresponding *monadic structural* universal by the mereological sum of the terms of the relation and then applies the partial-identity account of instantiation to the latter case.

later account of instantiation cannot support in a plausible manner the metaphysical necessity of laws of nature.

8.1.2 Dispositional Monistic Approaches

In opposition to categorical monists, dispositional monists argue that all laws are metaphysically necessary. For them, laws of nature 'flow' from the essence of natural properties or of natural kinds (of objects, fields and processes). Two distinct senses of metaphysical necessity are relevant here: According to the weak construal, laws are metaphysically necessary in the sense that they hold in all possible worlds in which the relevant nomically related natural properties exist. This weak metaphysical necessity is compatible with a kind of nomic contingency, since, if the nomically related natural properties are not necessary beings, then there are possible worlds in which they do not exist, and, therefore, the corresponding laws do not hold. A proponent of this view is Brian Ellis (2001), who claims that laws of nature express the essence of metaphysically contingent natural kinds.[4] In Ellis' view, nomic necessity is in a certain sense contingent for it concerns only those possible worlds belonging to the same natural kind with the actual world. The second interpretation yields a strong nomological necessity, according to which the actual laws hold in all metaphysically possible worlds. Alexander Bird (one of the main defenders of this view) claims (Bird (2007)) that, if P is a possible nomological property (if, in other words, it is a relatum of a nomic relation in at least one possible world), then it must exist in all metaphysically possible worlds. Given that the actual world belongs to the set of all possible worlds, it must include all possible natural properties. It seems, therefore, that if natural properties are necessary beings, then laws are metaphysically necessary in the strong sense.[5] For dispositional monists, the intuitions supporting the contingency of laws are (most often) explained away as presenting epistemic possibilities which, due to confusion, are regarded as genuine metaphysically possible scenarios.

---

[4] Brian Ellis is not a dispositional monist. I refer to him only as an example of a prominent metaphysician who embraces the weak nomological necessity.

[5] For a defence of strong necessitarianism, see also Bostock (2003).

Is there any room for nomic contingency in a dispositional monistic world? The answer seems to be yes. To illustrate the point, assume the strong metaphysical view that Dispositional Monism is metaphysically necessary; not only the actual, but also all possible worlds are inhabited by fundamental dispositional properties only. Does this assumption entail that the actual laws of nature are metaphysically necessary? Roberts (2010) disagrees; he starts by pointing out that the hypothesis that all natural fundamental properties have dispositional essences does not *by itself* entail the metaphysical necessity of laws. We have to further assume that the terms referring to the nomically related properties are *rigid*, that is, they refer to the same properties in all possible worlds. Relaxing this condition, one can show that those actual laws involving 'no-rigid' properties might not hold in other possible worlds. Let us see how this can happen. Roberts suggests that it is consistent with the thesis of Dispositional Realism (and consequently, with Dispositional Monism as well) to define non-rigidly the terms referring to nomically related properties via their causal/nomic roles. For instance, one may define a property P as the natural feature which in each possible world has a causal/nomic role akin to that of mass in the actual world. Hence, actually, P=mass, but in another world w (where, for instance, the gravitational constant might have a slightly different value) P=mass*. (Recall that the orthodox dispositional monist should hold that mass* cannot be identical with mass, since the actual causal/nomic role of the latter is essential to it.) Now, a kind of *transworld* law can be introduced which is functionally similar to Newton's gravitational law (with the actual gravitational constant), yet involves the property P instead of mass. In the actual world (where P=mass), the transworld law coincides with Newton's law and therefore holds.[6] In w, however, the transworld law does not hold since by definition it involves the *actual* gravitational constant. Hence, the transworld law is metaphysically contingent. Roberts claims that dispositional realists may adopt the view that all fundamental laws are of the

[6] I assume here, for the sake of the argument, that Newton's gravitational law holds in the actual world and is a fundamental law involving the fundamental property of mass. Obviously, Robert's suggestion can be applied to any property and law which physics might single out as fundamental.

transworld kind, despite their insistence on the 'traditional' form of laws with rigidly defined terms.

Another unorthodox proposal is due to Hendry and Rowbottom (2009). They challenge the 'orthodox' dispositionalist view which presupposes that the dispositional profile of each dispositional property is the same at least in all possible worlds in which the property in question exists. Their claim is that a dispositional realist may consistently hold a weaker thesis which allows a slight transworld variation in a property's dispositional profile.[7] According to their account (which they call Dispositional Contextualism), the possession of a dispositional property is associated with a *unique* set of actual *and possible* dispositions, where the latter can be slightly different from the former. Hence, although there is a kind of transworld *essence* associated with each dispositional property, laws of nature can be metaphysically contingent since, in each possible world, they 'flow' from a qualitatively different 'part' of this essence.

In the following sub-section, I am going to argue for the metaphysical contingency of laws of nature. Given my critical remarks on the possibility of nomic necessitarianism in a C-world, the most serious threat for nomic contingentism comes from the dispositional monistic camp. Hence, an available way to argue for the metaphysical contingency of laws is simply to follow one of the above-mentioned unorthodox accounts, which arguably show that nomic contingentism is consistent even with Dispositional Monism. I do not intend, however, to follow this course; I'll present instead an alternative argument which is based on the commitment to ConD/C.

8.1.3 An Argument for the Metaphysical Contingency of Laws

Consider the core claim of dispositional essentialism, according to which the causal roles that *any* natural property confers on its bearers are essential features which exclusively determine its *de re* modal representation (or, in the parlance of the transworld identity account, constitute its iden-

[7] Even if we grant that it is essential for any natural property to have a causal role, it does not follow that such a property should have the role it *actually* has in all possible worlds.

tity in every possible world in which it exists). In what follows I'll argue that a property theorist who rejects this claim is able to defend a kind of metaphysical contingency for laws of nature, *even if the relata of nomic relations are necessary beings*. Before embarking, however, on this task I must deal with two important objections. First, it might be objected that since the rejection of the essentialist claim is not motivated independently of the present task, the condition I place is utterly ad hoc. But this is not true, because we have reasons independent from the issue of nomic contingency that may convincingly ground that rejection. First, we may question the assertion that the essentialist claim holds for *all* fundamental natural features. As I have already argued, spatiotemporal relations (perhaps the best candidates for fundamental natural relations in the actual world) confer no causal powers on their bearers and so the essentialist claim does not hold for them. Furthermore, we may also question the validity of the claim in question even for those natural properties which dispositional essentialists themselves invoke as indisputably dispositional. Properties such as mass, electric charge and spin can be *de re* modally represented via a symmetries-based procedure, and so it is probably not the case that the powers that may confer on their bearers *exclusively* determine their *de re* modal representation.[8]

A second objection I wish to address is that the condition I proposed begs the question against the whole dispositional realist camp. If this is true, the upshot of the following argument is severely restricted. To meet this objection, I must point out that the core of the dispositional realist thesis is the acceptance of the existence of genuine, irreducible dispositional properties (and perhaps relations). Dispositional Realism is not necessarily related to the essentialist claim. Surely, a dispositional realist might follow this claim and reject my presupposition. But, given the above remarks concerning the essentialist claim, I don't think that following this course is a wise thing to do. Hence, in my view, the upshot of the following argument does not have a limited scope; for, the majority of property theorists reject or should reject the essentialist claim and so accept the condition which my argument relies on.

[8] Psillos (2006) mentions the symmetries-based manner of identification in the context of an argument against the intelligibility of a powers-based ontology. For details, see also Livanios (2010).

Let me now turn to the argument itself. Since categorical monists must endorse the metaphysical contingency of laws in order to render their view maximally plausible (recall my comments on Fales and Armstrong in Sect. 8.1.1), the aim of the argument is to show that dispositional realists (whether they are dispositional monists or dualists) can also be nomic contingentists. As I have already remarked, this conclusion can only be established provided that dispositional realists reject the essentialist claim. But how can they do that? In my view, a reasonable way to do that is to embrace ConD/C. Dispositional realists can hold this view because, independently of what they believe about the *actual* metaphysical character of any natural feature (whether it actually is genuinely dispositional or not), they are not committed to the view that this property or relation is *necessarily* such. In other words, they are not committed to the view that in all possible worlds where the feature in question exists, it is, for instance, dispositional. By endorsing ConD/C, dispositional realists are forced to reject the troublesome essentialist claim while at the same time retaining the core of their view. They *have to* reject the essentialist claim because they cannot, for instance, hold that the identity of a fundamental natural property in a world in which it is categorical is exclusively constituted by the causal roles that the property in question confers on its bearers in possible worlds in which it is dispositional. According to dispositional essentialists, causal roles are supposed to be essential features of natural dispositional properties. But they cannot be if the latter can exist in other worlds as categorical, because in that case they cannot possess essential features related to their causal roles. Armed with the premise of ConD/C, I can now show how dispositional realists may defend the view that laws of nature are metaphysically contingent. Here is my argument:

(1) All fundamental natural properties are necessary beings.

The first premise entails that all actual fundamental natural properties exist in all possible worlds. Its role is to ensure that the conclusion of the argument does not concern the well-known 'thin' metaphysical contingency of laws due to the alleged contingent existence of nomically related fundamental properties.

(2) Two fundamental natural properties P and Q, which are dispositional in a possible world w, and the nomic relation L such that the state of affairs L(P, Q) is a law of nature.

The second premise sets up a scenario that prima facie constitutes *the* favourite case for nomic metaphysical necessity. Dispositional Realism (as most contemporary philosophers understand it) dictates that if *all* nomically related natural properties are dispositional, then laws must be metaphysically necessary.

(3) L is an internal nomic relation in w.

Premise (3) follows from premise (2) and what I have said in Sect. 7.1.

(4) P and Q are contingently dispositional.

Premise (4) assumes the truth of ConD/C (for a defence, see Chap. 6)

(5) There are possible worlds in which both P and Q exist and are categorical.

It follows directly from premise (4).

(6) In the worlds mentioned in premise (5), there is no ontological warranty that L exists.

Premise (6) follows from the categorical character of the relata of L. Notice that the possible non-existence of L in the aforementioned worlds has nothing to do, in our case, with the existence of its relata.[9]

∴ (a) There are possible worlds in which L does not exist, though its relata do. Hence, the law L(P, Q) is metaphysically contingent.

The conclusion I have reached is not altered if we assume that L has more than two dispositional relata. In any case, the restriction to nomic relations

[9] Here I assume that Armstrong's partial-identity account is false, at least as far as the case of laws is concerned. For the relevant arguments, see Sect. 8.1.1.

with only two relata is not substantial and was made for reasons of convenience. But most importantly for our case, neither is it altered if we assume that L has at least one categorical relatum. For, I can reformulate the scenario of my argument and appeal to a possible world where all dispositional terms of L turn out to be categorical, and where the initially categorical relatum retains its categorical 'nature'. So, to conclude: (a) given that for dispositional monists the case under consideration in my argument exhausts *all* possibilities for nomic relations, the result I have reached clearly shows that they can accept a kind of nomic contingency provided they acknowledge the possibility of ConD/C (and, of course, amend their view accordingly). (b) The fact that this result also holds in the modified case of a nomic relation with at least one categorical relatum shows that property dualists too can accept nomic contingency under the aforementioned condition. Hence, given that nomic contingency is most plausibly a main thesis of Categorical Monism as well, the arguments of this sub-section demonstrate that all property theorists can be nomic contingentists provided they are dispositional contingentists.

I would like to conclude this sub-section by pointing out a further conclusion that follows from the above argument. Given that L has in some possible worlds categorical relata, it cannot be *necessarily* an internal relation (provided, of course, that Armstrong's account which arguably leads to the view that all nomic relations are internal is false). I am sure that a number of philosophers may find this view absurd, mainly because they think that metaphysical features such as the internal/external character of nomic relations *must* be metaphysically necessary. Yet, it is the assumption that internality and externality are *essential* features of nomic relations that supports the belief that they are metaphysically necessary as well. Refuting this assumption, we can construe the kind of metaphysical underdetermination[10] emerging from the existence of different metaphysical views as a

---

[10] To avoid misunderstandings, this is not the kind of metaphysical underdetermination which ontic structural realists appeal to for motivating their view. According to this often-discussed underdetermination, distinct metaphysical packages are all consistent with a fundamental theory (for instance, individualist and non-individualist accounts of particles' nature are both consistent with quantum theory). This fact implies that theory cannot tell us what the world is like as far as the nature of its objects is concerned (for more details about the varieties of underdetermination and possible reactions to them, see French (2011)). In our case, we do not have a mature scientific theory that posits entities the nature of which is metaphysically underdetermined; rather, it is a belief shared by a number of metaphysical accounts which presupposes the existence of entities characterised by a metaphysically underdetermined feature.

sign of metaphysical contingency of the above features of nomic relations. Hence, regardless of which metaphysical account is *actually* true, the other accounts on offer present different meaningful metaphysical images of how the world could have been. To put it differently, different metaphysical accounts describe the features (internality or externality) which actual nomic relations could possess in possible worlds that differ from the actual one with respect to the kinds of natural properties that they have.

8.2 Hybrid Nomic Relations

In this section I'll examine the nature of mixed nomic relations (especially those that neither involve fundamental constants nor are involved themselves in conservation laws). Is it reasonable to regard these nomic relations as internal? In order to answer this question, I'll focus on a specific *kind* of mixed nomic relations the categorical relata of which are spatiotemporal relations. This choice has the advantage of considering fundamental features which all but dispositional monists agree are categorical. It might, nevertheless, raise sceptical doubts about the *universality* of the results to follow. In any case, I would rest content with the modest conclusion that there is a *species* of mixed nomic relations with interesting (as I'll argue) metaphysical features.

Given all that, let us now suppose, for the sake of illustration, that the actual world is an M-world and the nomic relation involved in Coulomb's law is a fundamental one. In order to sidestep the (irrelevant for the issue at hand) difficulties concerning the nature of this law *qua functional* one, let us focus on its 'determinate' form:

$$F = kq_1q_2 / r^2 \quad (*)$$

where q_1 and q_2 are *specific* values of electric charge, r is the spatial distance between specific instances of these charges, F is the force one charge-instance exerts on the other and k is a non-fundamental constant. Formula (*) expresses a second-order fact, one of the constituents of which is the nomic relation relating charges and the spatial distance between their instances. Following the property-dualistic view, we may construe the former as fundamental dispositional properties and the latter as a categorical relation. From this perspective, Coulomb's law involves a mixed nomic relation. The question then is, Is it plausible to think of this mixed nomic relation as an internal one? One way to do this is to claim that internal relations may supervene on the natures of only *some* of their terms. One might suggest, for instance, that the nomic relation supervenes on the dispositional nature of charges and not on the categorical spatial distance. This view avoids any difficulty concerning what relationship can the categorical 'nature' of distance (if any) have with the specific nomic relation. Yet, it faces an obvious difficulty; it seems clear that the nomic relation under consideration depends *in some sense* on r. The question is what kind of dependence this is. Those philosophers who aim to construe all nomic relations as internal would most probably claim that the *fundamental* features of the world are only those on which nomic relations supervene, that is, the natures of dispositional properties. The spatial relation (*qua* categorical) is not a fundamental feature of the world and the dependency of the Coulomb force on it should be 'absorbed' by the natures of the charges. The suggestion would be this: It is of the (intrinsic) nature of charges q_1 and q_2 that, when their instances are a distance r apart, they attract or repel each other by the specific force F. From this perspective, the ontological status of spatiotemporal relations is downgraded. They become a kind of 'intrinsic' (to the dispositional charges) factor that determines only the *strength* of the Coulomb force (and similarly for other nomic relations, such as Newton's gravitation law, which involve manifestations-as-forces of other dispositional properties). My response to that is based on the assumption that a genuine M-world should 'possess' fundamental categorical features which are not ontologically dependent on the dispositional ones. For instance, the actual world *could have been* a genuine M-world (in fact, that is what supports the possibility which we now talk about) since GR (which is, till now, the most successful and mature spacetime theory) suggests under a defensible interpretation that spacetime actually exists and is at least as fundamental as matter fields.[11] Given that the already suggested 'absorbing'

[11] It might be claimed that some contemporary physical theories do not support such an interpretation since, according to them, spacetime is an *emergent* feature of our world and does not belong to its fundamental ontological furniture. My response is that, though I am not sympathetic to science-

interpretation refutes the ontologically independent and robust existence of categorical features, it grounds an implausible view concerning the relation between fundamental dispositional and categorical features in an M-world.

Another way to uphold the view that the nomic relation in question is internal due to its supervenience on some of its terms is to claim[12] that it is of the nature of the spatial distance r that when two charges q_1 and q_2 are r apart they attract or repel each other with a force F. The advocate of this view must admit that a spatiotemporal relation is a *special* kind of categorical relation the nature of which is not exhausted by the definitional characteristics of the nature of categorical features, at least as traditionally conceived. As I'll argue below, though this view can be defended, there is another, more tenable view that one might hold. In any case, and regardless of the success of all the above remarks, there are some general thoughts that might question the tenability of the original suggestion. For instance, how can an internal relation supervene on the nature of only some of its terms? What is it that determines which terms it is supervening on?[13] Paradigmatic internal relations (such as resemblance) supervene on the nature of *all* of their terms. Is it not an ad hoc claim, therefore, to suggest otherwise in the very special case of the nomic relations under consideration? If, on the other hand, it is actually the case that some nomic relations are metaphysically special, why not describe them by using an entire new metaphysical category?

If all the above points persuade us that it is rather implausible to assume that some nomic relations are internal because they supervene on some of their relata, we may follow the alternative course and claim that the nomic relation of our example does supervene on the nature of *all* of its terms (dispositional and categorical alike). Is this suggestion a reasonable one? I think not, because the nature of categorical terms is by definition meta-

insensitive metaphysics, I think it is wiser not to base any metaphysical views on the findings (or, most often, speculations) of immature and (in some cases) under-developed theories.

[12] See, for instance, Mumford's (2004, 188) speculation.

[13] Perhaps it is a brute fact that it is of the essence of some terms (as opposed to the essence of others) to belong to the supervenience base. Acknowledging this, however, does not increase the plausibility of the suggestion under discussion. The relevant point here is that in denying that the internal relation supervenes on *all* of its terms, one has no cogent reasons to choose some of the terms as belonging to the supervenience base.

physically unrelated to their nomic roles. Hence, they cannot determine any nomic relations involving them. (This fact, of course, does not preclude them of determining other, non-nomic, relations holding among them.) It might be objected that this conclusion follows only by using too *narrow* a meaning for the nomic roles of spatiotemporal relations. Perhaps, as I have tentatively suggested above, spatiotemporal relations are not 'standard' categorical features, and so their nature is not exhausted with the definitional characteristics of the nature of categorical features as traditionally conceived. It is possible to claim (taking our cue from a relevant suggestion of Ellis (2001)) that it is of their nature to set the 'stage' for causal powers to operate, and that it is *this* fact which constitutes (at least a part of) their nomic role. In that case, it is simply not true that the nature of spatiotemporal relations is metaphysically *unrelated* to their nomic roles. Is there a problem with this proposal? My reluctance to accept this view stems from the fact that, according to this account, spatiotemporal relations cannot be plausibly construed either as 'standard' dispositional or as 'standard' categorical features. Hence, to appeal to this view in order to defend the claim that the nomic relations in question are internal, we need to acknowledge that the categorical/dispositional distinction is *not* exhaustive. Perhaps this is true and we have to accept the existence of a *third* kind of natural properties and relations, one in which spatiotemporal relations belong. I think, however, that this view, though possible, is not to be preferred, because there is an alternative, more reasonable, one. To be more precise, we may insist that the dispositional/categorical distinction *is* exhaustive and meet the difficulty concerning the metaphysical character of the nomic relations under consideration by claiming that the internal/external distinction is *not* exhaustive. As we'll soon see, this alternative view has a precedent in the literature in the work of Keith Campbell. Campbell's third characterisation (which proves the non-exhaustiveness of the external/internal distinction) can be easily understood because it is presented through an example of a property of *everyday* life which has nothing to do with nomic and spatiotemporal relations. And it seems more reasonable (and non-ad hoc) to place the nomic relations we examine in a new metaphysical category that includes other entities than to claim the existence of an extra category of properties and relations which, as far as we know, includes spatiotemporal relations *only*. I conclude that there is no plausible way to argue that the actual nomic relations which relate both kinds of properties and relations (dispositional and categorical) are internal.

My suggestion that the proper characterisation of some nomic relations compels us to reject the exhaustiveness of the external/internal dichotomy cannot even get off the ground if we have cogent reasons to think of these nomic relations as external. I can show, however, that not only we have no such reasons, but, on the contrary, we should acknowledge the fact that these relations *cannot* be external. To illustrate the impossibility of the aforementioned nomic relations being external in the presence of dispositional relata, consider again the case of the nomic relation involved in (the determinate form of) Coulomb's law. Can we construe this nomic relation as external? Recall that an external relation does not, by definition, supervene on the natures of its relata. So, for the nomic relation of our example to be external, we must assume that it does not supervene on the nature of *any* of its relata. But this cannot be true. Nomic relations, such as the one in our example, do not supervene on the natures of their categorical relata (i.e., the spatial relation) because the latter have a nature that is metaphysically unrelated to their nomic roles.[14] But it cannot be plausibly said that neither do they supervene on the nature of their dispositional terms (i.e., the charges); for, due to the dispositional nature of charges (a nature intimately related to their nomic role as it is expressed in Coulomb's law), the nomic relation must supervene on their nature. Hence, the nomic relations of the actual world which involve both dispositional and categorical terms cannot be external.

The upshot of the above discussion (namely, that mixed nomic relations are *neither* internal *nor* external) leaves the crucial question unanswered: What is the nature of those mixed nomic relations? In order to present my account, it would be useful to begin by considering a specific example of a relation that Keith Campbell (1990, 112) presents. In particular, Campbell (while examining the prospects of the doctrine of *foundationism*[15]) discusses the case of the relation 'is sweeter wine than' which

[14] The objection that this claim is supported by a limited conception of the nomic role of spatiotemporal relations has been already met. Similar remarks apply in the present case.

[15] The doctrine according to which there are no relational differences without differences in monadic properties.

holds between two wines in the actual world and has a feature that is most interesting for our investigation. Let us assume that a specific sweetness belongs to the nature of each wine. Then it is reasonable to claim that the aforementioned relation supervenes on the nature (i.e., the specific sweetness) of its terms, but, nevertheless, is *existentially dependent* on the presence of living beings endowed with sense organs sensitive to sweetness. The reason for the latter dependence is that the very nature of its terms depends on the presence of appropriately equipped beings, an external factor to the relation and its terms. It is absurd to attribute a specific sweetness to a wine in the absence of beings appropriately equipped.[16] Sweetness is a kind of 'extrinsic' property of wine (in the sense that its existence depends on the presence of an external factor), but, nevertheless, does belong to its nature. In the case of the relation 'is sweeter wine than'" the presence of the external factor is implicit. The relation does not supervene on this external factor, but it depends on it for its existence. It is an *external founded* relation, as Campbell calls it. Returning now to our case, I suggest that those (mixed) nomic relations that relate both dispositional and categorical properties and relations belong in the category of externally founded relations. More precisely, they belong in a subcategory of externally founded relations that includes relations in which the external founding factor is *explicitly* present as a term. To illustrate, let us consider once again Coulomb's law in its 'determinate' form. The corresponding nomic relation supervenes on the nature of some of its terms, namely, the dispositional charges. Nevertheless, the relation exists and the supervenience holds only in those worlds where spatiotemporal relations exist, because the nature of charges themselves involves, as an external factor, the presence of spatiotemporal relations.[17] Of course, we must be

[16] One possible objection is that we may implicitly attribute a specific sweetness to a wine by identifying the independent-from-sentient-beings basis in the wine for its sweetness with its sugar content. Yet, I think, even in that case, the crucial point of Campbell's example still holds. Recall that what Campbell wants to show is that the nature of the relata of the relation in question is dependent on external factors. As long as it is sweetness (and not its basis in the wine) that is part of the nature of the relatum of Campbell's relation, it certainly needs for its existence an external factor (the taste of a sentient being).

[17] To avoid misunderstandings, I have to remind the reader that, throughout this book, talking about the 'nature' of properties does not necessarily have anything to do with essential (or, simply, metaphysically necessary) higher-order features. My 'loose' use of the term allows me, for instance, to refer to the 'dispositional nature' of a fundamental property, though I think that dispositionality

cautious in the interpretation of this kind of involvement. My intention is not to claim that the intrinsic nature of each charge (consisting at least partially of its causal powers) 'absorbs' the spatiotemporal dependency of the force exerted on the other charge. Rather, what I mean is that it is of the nature of each charge to need categorical spatiotemporal relations setting the 'stage' for its causal powers to be operative (see also the relevant suggestions of Brian Ellis (2001, §3.12; 2009, Ch.5). Hence, the nomic relation under examination is an instance of an external founded relation that explicitly incorporates (as a term) the external factor (spatiotemporal relation) on which it existentially depends. It is convenient to give a name to the sub-category which includes such nomic relations. I suggest calling it the sub-category of *hybrid* relations. According to my account, therefore, some nomic relations are hybrid. Their hybrid character demonstrates how a relation can supervene on the natures of only some of its terms while simultaneously existentially depending on the other(s). The *fundamental* character of the categorical terms–external factors is guaranteed because their role is not 'absorbed' in the intrinsic nature of the dispositional terms. On the contrary, they constitute those fundamental features the existence of which existentially grounds the 'extrinsic' nature of dispositional terms.

It might be objected (recall earlier remarks on that issue) that those hybrid nomic relations can also be construed as internal, provided that we decide to think of spatiotemporal relations as neither dispositional nor categorical but as belonging in a third metaphysical category. Yet, it seems that following this possible course is an ad hoc move because, as far as I can see, there is no independent reason to introduce such an extra metaphysical category of natural features besides the case under consideration. Contrary to the above-mentioned approach, the appeal to the sub-category of hybrid relations is not ad hoc, since the encompassing category of external founded relations is already acknowledged as the proper metaphysical category accommodating the case of other relations which have nothing to do with spatiotemporal and nomic relations. This

is a metaphysically contingent feature of natural properties. Consequently, to be precise, the proper range of worlds in which supervenience of a mixed nomic relation on the nature of its dispositional terms holds, includes only those possible worlds where *both* spatiotemporal relations exist and the actual dispositional terms are *still* dispositional.

latter fact vindicates and supports my preference of the hybrid-relations metaphysical account. Finally, I must point out that hybridity (just like internality and externality) is a metaphysically *contingent* feature. In my view, this metaphysical contingency is not a *brute* modal fact. Rather, it can be construed as 'flowing' from the contingent dispositionality/categoricality of those fundamental features which are the relata of nomic relations. For instance, there are possible worlds in which all dispositional relata of a hybrid nomic relation are categorical and, consequently, the relation is no longer hybrid but external.

It is now time to recap the conclusions we have reached thus far in this chapter. Some nomic relations are neither external nor external but externally founded (hybrid), but not necessarily so, since there are possible worlds in which they are internal or external. Finally, all fundamental laws of nature are metaphysically contingent.

8.3 The Role of Hybrid Nomic Relations in an M-World

In Sects. 7.1, 7.2 and 7.3, I have implicitly argued for a selective nomic realism in M-worlds. My strategy has been to undermine DEAL, a version of which (the one holding that spatiotemporal factors are 'absorbed' in the dispositional essences) articulates the worst scenario for the genuine existence of mixed nomic relations in M-worlds. I have concentrated on the cases of conservation laws (and the related symmetries) and fundamental constants appearing in laws, because I think that they provide clear reasons to reject DEAL as inadequate.

Arguing, however, for the genuine existence of mixed nomic relations in M-worlds does not answer another crucial question: What is the role of mixed nomic relations in an M-world? How do they earn their keep? Recall that dispositional essentialists insist that the existence of genuine fundamental dispositional properties ontologically downgrades laws of nature. Non-eliminativists believe that laws supervene on the fundamental dispositional properties and merely express the internal relations among them (due to their intrinsic nature). Eliminativists take a step further and claim that dispositional properties can play all roles that laws are supposed to play and so render them ontologically redundant. Given, then, that (a) M-worlds are (by definition) populated by genuine fundamental dispositional properties which are relata of mixed relations, and (b) both dispositional monists and dualists ontologically downgrade all nomic relations with dispositional relata, what is the role of mixed nomic relations in an M-world? Providing an answer to this question will give us a clue about how to answer another important question: Can we believe in the existence of ontologically robust laws in an M-world? These two questions are related to each other, since it is assumed that laws which have no role to play are at best ontologically second-rate entities.

Now, one might challenge the dispositional realists' claim by suggesting that laws may earn their keep by filling explanatory roles that dispositional properties cannot fill. Consider, for instance, Psillos' (2008, 188) challenges to the dispositional realist who rejects the existence of robust laws.[18] First, what is the ontological explanation of the fact that some dispositional properties come always together (as features of entities of certain kinds) and others do not? Second, what is the ontological explanation of the fact that certain capacities or dispositions (such as the capacity of a massive object to move faster than light) are absent in the actual world? Third, what is it that ontologically distinguishes genuine dispositional properties from mere Cambridge dispositional ones? In each of the above cases laws can provide a reasonable explanation; it might be that laws compel some dispositional properties to be co-instantiated at certain objects; furthermore, the absence of a specific disposition might also be a consequence of a specific law; finally, genuine dispositional properties (in opposition to mere Cambridge ones) appear as relata in nomic relations. Nevertheless, the dispositional essentialist may meet these challenges either by providing alternative explanations based on the dispositional natures of properties or by insisting that at least some of the disputed facts need no explanation. For instance, she may claim that it is a *brute* fact concerning any possible world that only *some* of its

[18] Here, I focus on some ontological points. Psillos (2008) also presents some *epistemic* difficulties that emerge in a dispositional-realistic context and can be met by assuming the genuine existence of laws.

dispositional properties are genuine fundamental features. Furthermore, she may insist that it is of the *nature* of fundamental properties that they do *not* confer on their bearers certain dispositions. In each case one has to weigh the merits of each ontological explanation; the superiority of the laws-based explanation is not given in advance. Finally, there could also be other explanations based neither on dispositional natures nor on laws of nature. In fact this is the case with the first challenge, where the 'sociability' (to borrow a term from Chakravartty (2007) and (2007a)) of certain dispositional properties described by the Standard Model of elementary particles can be explained[19] in terms of symmetry principles (consider, for instance, the explanatory role of the permutation group of transformations in the classification of elementary particles as fermions and bosons—see French (2013)).

Another argument[20] for the indispensability of genuine laws in a world inhabited by genuine dispositional properties can be advanced if we consider the most plausible interpretation of the nomic role (in such a world) which is consistent with actual scientific practice. Regularities of observable events, *as precisely described by nomic relations,* are very rare due to the ubiquitous existence of interference factors (such as antidotes). Laws describe precisely the behaviour of physical systems in rare, experimentally created, 'shielded' conditions which most often are physically unrealised. Nevertheless, scientific findings rely on this restrictive nomic role, as the success of the analytic method and the method of abstraction clearly shows. For some dispositional realists, the most cogent explanation of this fact is that laws (better, nomic statements) do not state how physical systems of a certain kind *actually* behave, but rather how they *would* behave in precisely specified—albeit mostly physically unrealised—circumstances. According to this view, laws ascribe to objects *stable* (i.e., invariant under all changes of conditions) *dispositions* for particular behaviours associated with the relevant dispositional properties instantiated by objects in question. Now, if we assume that dispositional properties have typical manifestations which are not realised in the observable

[19] Of course, the explanation is ontological provided that we also endorse an ontological interpretation of the relevant symmetries.

[20] This argument (not exactly in the present form) has been advanced in Corry (2011).

sequence of events due to the presence of interference factors, we have to posit unobservable entities which *are* these typical manifestations and, crucially, we *need* laws to tell us how these entities interact/combine with interferences to 'produce' the actually realised observable behaviour. To illustrate, consider the best candidates for these manifestation-entities, component forces. The actual behaviour of objects on which various forces act can only be described/explained by appeal to a law of composition of these forces (in order to get the total force) and Newton's second law (in order to associate the observable effect, that is the acceleration of objects, with the action of forces). Dispositional realists might try to avoid the upshot of this argument by claiming that the posited laws can also be understood as 'flowing' from the dispositional natures of the component forces themselves. But that would not do. Consider, for instance, Newton's law; applied with respect to each component force it cannot be deemed a description of what actually happens because of the presence of other component forces. If dispositional realists claim that the said law describes what *would* happen in the absence of other forces, an infinite regress looms. Because they have to once again posit new entities which *are* this time the stable typical manifestations of component forces, and they crucially need *laws* to tell them how these new entities interact/combine with interferences to 'produce' the actually realised observable behaviour. The infinite regress is avoided provided that Newton's law is applied only with respect to the *total* force acting on each object. But this presupposes that the law of composition of forces can be cast in terms of the dispositional nature of forces themselves. But if one is willing to 'absorb' into the nature of forces the results of their combination with other (similar in kind) entities which act as interferences, she should not do this at the level of forces but rather at the level of the dispositional properties themselves.

In my view, the above argument fails because it mistakenly requires that the laws in question must 'flow' from the dispositional nature of *forces*. Dispositional realists, however, can plausibly claim that these laws express the dispositional nature of properties themselves. One way to do this is to associate the 'nature' of each fundamental dispositional property with a variety of 'interference' manifestations presented as total observable effects. The availability of this move shows that the argument under consideration cannot prove the indispensability of laws in a world inhabited by genuine dispositional properties.

Taking my cue from Katzav's (2005) remarks, we may formulate another argument for the existence of some laws the truthmakers of which cannot be the dispositions that fundamental dispositional properties confer on their bearers. Katzav describes a scenario according to which there are in a possible world two dispositions which, if possessed in the proper conditions (not logically mutually exclusive), will yield *contradictory* manifestation events. Dispositional realists usually think that the compatibility of the *actual* fundamental dispositional properties can be plausibly construed as a *brute* fact. But this would not do in Katzav's problematic case. One may then naturally wonder how dispositional realists can secure the consistency of *possible* worlds inhabited by dispositional properties. Dispositional realists' favourite answer is that it is the dispositional nature of the actual fundamental properties that guarantees the impossibility of the imagined scenario. Yet, Katzav's crucial point (which grounds the possibility of the suggested scenario) is that there could be no law of nature 'flowing' from the nature of the *actual* fundamental dispositional properties that would determine which non-actually existent dispositions are possible, and hence, that there could be no such law that would exclude the coexistence of dispositions with contradictory manifestations in a possible world. The upshot is that in any world containing dispositional properties (be it a D-world or an M-world) there are ontologically robust laws which are not supervenient on the natures of the dispositional properties.

Katzav's intention is, however, to challenge dispositional ontology, not to argue for the existence of laws. The crucial difference from my reading of his remarks is that he thinks (following orthodoxy) that the emerging difficulty can only be avoided by assuming the *necessary* compatibility of fundamental dispositions. In this sense, he rightly seeks *necessary* laws of nature that can guarantee the desired result. Yet, he can find no such laws; both Humean regularity-based laws and DTA laws are metaphysically contingent, while 'dispositional' laws (grounded in the nature of dispositional properties), which are arguably necessary, cannot do the job, as he insists, hence the challenge facing dispositional realists to defend the consistency of their ontology. Though I think that the actual world is a C-world, I feel the burden to respond to Katzav's challenge because I believe that both D-worlds and M-worlds are metaphysically *possible*. My response, then, is this: Since, in my view, fundamental dispositional properties are *contingently* dispositional, we do *not* need to assume their necessary compatibility to avoid the possibility of inconsistency. We just need their compatibility in all those worlds in which they are still dispositional, and not in all possible worlds in which they (or their counterparts) exist. But this can be ensured through metaphysically *contingent* laws, either of the Humean or the DTA kind.

Though I am sympathetic to the above kind of argument, I do not think that we can use it to defend the genuine existence of mixed nomic relations. As far as I can see, we have no reason to think of Katzav's laws as involving mixed nomic relations. So we have to return to our initial question and search for an indispensable role that mixed nomic relations can fill in an M-world. Though I do not have a general argument to support such a role in *each* possible M-world, I do think that mixed nomic relations are indispensable at least in those M-worlds with fundamental categorical features which are ontologically independent (of fundamental dispositional features) spatiotemporal relations. The spacetime of these M-worlds may be correctly described by GR or any other theory which suggests that spatiotemporal relations are genuinely existing fundamental features of those worlds. To support my claim, recall that, while analysing the 'nature' of hybrid nomic relations, I've remarked that the latter are existentially dependent on the categorical spatiotemporal relations and, by definition, involve dispositional properties as well. Motivated by this fact, I now suggest that one may plausibly conceive of the hybrid nomic relations (at least the ones involving spatiotemporal categorical features) as providing the proper ontological 'link' between the dispositional and the categorical features of an M-world. In other words, given that in an M-world there exist two distinct *independent* kinds of fundamental feature, hybrid nomic relations are needed to provide the metaphysical 'glue' that unites them in a coherent ontological whole.

One might, of course, object that the above proposal has a *limited* scope, since there surely exist metaphysically possible M-worlds which either have no spatiotemporal relations, or, though they have, these are not categorical (in that case, of course, some *other* fundamental features must be categorical). I agree, and that is why I emphasise that I do not offer here a suggestion concerning the role of mixed nomic relations in *all* M-worlds. Notice, however, that to the extent that a *genuine* M-world ought to 'possess' fundamental categorical features existing independently of the dispositional ones, my proposal cannot be threatened on the grounds that fundamental dispositional properties may ontologically 'involve' somehow categorical spatiotemporal relations. As I argued in the preceding section, such an assumption, by being a version of the 'absorbing' strategy of categorical features into dispositional natures, clearly suggests that categorical terms of hybrid nomic relations are ontologically non-fundamental, because their role is entirely 'absorbed' in the intrinsic 'nature' of dispositional terms.

9

Concluding Remarks

When I began to write this book, I had two main aims; the first, more specific one, was to defend a kind of sui generis Categorical Monism for the fundamental features of our world and explore the consequences of this view for the metaphysics of laws of nature. To this end, I argued against Dispositional Monism by showing that spatiotemporal metric relations are purely categorical. I also rejected Identity Theory, mainly because of the lack of an adequate explanation of the 'surprising' identity grounding it, and Neutral Monism, due to its concomitant unpalatable agnostic stance towards the proper nature of elements supporting the different metaphysical worldviews. Finally, I defended Categorical Monism itself by offering a novel argument from scientific practice. A crucial point emerged during the discussion is that the science-sensitive metaphysical methodology commits us to a kind of metaphysical *contingentism* concerning the views we argue for. Hence, the sui generis character of Categorical Monism defended here, which is *contingently* and *not necessarily* true. Based on the metaphysically contingent Categorical Monism (in particular, the corollary that dispositionality and categoricality are metaphysically contingent features of the fundamental properties),

V. Livanios, *Science in Metaphysics*,
DOI 10.1007/978-3-319-41291-7_9

I argued for a robust kind of metaphysical contingency for the actual laws of nature and a selective nomic realism in all possible worlds. Finally, a third category of relations (beyond the internal and external kinds) has been introduced, the *hybrid* one, in which some mixed nomic relations (in dualistic possible worlds allowed by the metaphysical contingency of Categorical Monism) fall.

Every philosophical work is characterised by certain omissions and lack of sufficient support for some of its claims. Of course, this book is no exception to this rule. Some of its weaknesses are due to lack of space. That is, for instance, the case with the insufficient defence of my preferred ontological criterion of the dispositional/categorical distinction and the commitment to a DTA-like theory of laws. That is also the case with the total neglect of the Double-Aspect (aka Dual-Sided) Theory of properties (though the latter is most probably justified, given that almost no one defends it nowadays). There are, however, other 'restrictions' in this work not falling to the above-mentioned case. Such an obvious 'limitation' concerns the range of natural features I focus on. Recall that (a) I insisted that the proper context for the metaphysical discussion related to the debate between categoricalists and dispositionalists is the one of the fundamental features of the actual world, and (b) concentrated on the spatiotemporal relations and the various charges involved in the fundamental interactions. Even if we grant the first claim, one might wonder why we should 'restrict' the discussion on the features examined here while neglecting other properties contemporary physics also posit as characterising elementary physical entities. Examples of such properties are position, momentum, angular momentum, energy, and so on, which take values that depend on the state of the physical entity (or the system). A number of philosophers might conclude from this 'restriction' that the view defended here has a narrower scope than that being assumed; and given that, one is very close to take a step further and think that my arguments (even if they are sound) allow a kind of Property-Dualism as the true metaphysical account of the actual fundamental features.

Of course, the verdict about the proper choice of the set of features to be scrutinised in this context depends heavily on the *appropriate* notion of fundamentality. One might appeal to the various interpretations appearing at the metaphysical literature (for instance, the suggestion that the fundamental features are elements of the minimal collection of entities which either individually or collectively provide a complete metaphysical ground for all other features), but it is not clear (at least to me) that following this strategy would help us to specify which natural features are fundamental and which are not. Things might be even worse, since it is possible that there are *no* fundamental features even in the 'weak' sense of those characteristics ascribed to the entities of the fundamental level; for, there is a possibility of an infinite chain of descending levels of ontological dependence (see Morganti (2014); see also McKenzie (2011) for some anti-fundamentalist conclusions based on naturalistic grounds). Nevertheless, the appeal to physics itself might provide some hints. It is not unreasonable, for instance, to claim that the fundamental features are exactly those which play an indispensable role in the theoretical explanation of the fundamental interactions of our world. That claim vindicates my choice, since the spatiotemporal relations and the various charges play such a role in the contemporary theoretical context. Furthermore, the fact that the existential status of other physical properties and relations depends on the interpretation of Quantum Mechanics might be seen as vindicating the 'restriction' to charges and spatiotemporal relations.

For some philosophers, the above remarks are far from convincing; even these, however, may charitably interpret my 'restriction' as based on a legitimately adopted agnostic stance. To the very least, and given that the dispositional/categorical debate has so far been almost exclusively concentrated on the natural features I focus on, one might see the arguments of this book as contributing to a specific *phase* of the dialectics of the ongoing debate. Under that perspective, she might consider the 'limitations' of the present work as offering the proper *motivation* for further work in order for the view developed here to be more firmly defended. For instance, given that, under certain assumptions, the categorical/dispositional debate makes sense even in a purely structuralistic context (in which, *fundamentally*, there are only structures), a structuralistic 'extension' of the debate could be a fertile ground for novel arguments fostering its further development (in fact, some of the arguments of this book might also apply *mutatis mutandis* to this context (see, for instance, Livanios (2014)). Closely related to this project is the topic concerning the proper interpretation of fundamental physical symmetries. I am sympathetic to a (selective) ontological interpretation, but surely a lot of effort should be devoted in order to have an adequate metaphysical account of (all?) fundamental symmetries and clarify their metaphysical role within the overall structure of the actual world.

This book, however, is *also* intended to be an extensive case study demonstrating what it could possibly mean to follow the approach to metaphysical problems which I call *Science in Metaphysics*. A crucial characteristic of this approach is that the role of science in metaphysical debates is by no means *exhausted* to the function of being the (ultimate?) arbiter for the truth or falsity of rival metaphysical views. The relationship between science and metaphysics is far more complex than we might initially think and the project of analysing it is still at its embryonic stage. In my view, the most fruitful way to proceed is to engage in particular metaphysical studies and highlight, case by case, threads of the multifaceted web of relations between metaphysics and science. Hopefully, the study at hand will be deemed successful in doing exactly that.

References

Adams, F. (2008). Stars in Other Universes: Stellar Structure with Different Fundamental Constants. *Journal of Cosmology and Astroparticle Physics* 08, 0810.

Anderson, J.L. (1967). *Principles of Relativity Physics*. New York/London: Academic Press.

Armstrong, D.M. (1978). *Universals and Scientific Realism: Nominalism and Realism*. Cambridge: Cambridge University Press.

Armstrong, D.M. (1978a). *Universals and Scientific Realism: A Theory of Universals*. Cambridge: Cambridge University Press.

Armstrong, D.M. (1983). *What Is a Law of Nature?* Cambridge: Cambridge University Press.

Armstrong, D.M. (1989). *A Combinatorial Theory of Possibility*. Cambridge: Cambridge University Press.

Armstrong, D.M. (1996). Reply to Martin. In Crane, T. (Ed.) *Dispositions: A Debate*. London/New York: Routledge, 88–104.

Armstrong, D.M. (1997). *A World of States of Affairs*. Cambridge: Cambridge University Press.

Armstrong, D.M. (2004). *Truth and Truthmakers*. Cambridge: Cambridge University Press.

Armstrong, D.M. (2004a). How Do Particulars Stand to Universals? In Zimmerman, D.W. (Ed.) *Oxford Studies in Metaphysics*, Vol. 1. Oxford: Clarendon Press, 139–154.

V. Livanios, *Science in Metaphysics*,
DOI 10.1007/978-3-319-41291-7

Armstrong, D.M. (2006). Particulars Have Their Properties of Necessity. In Strawson, P.F. & Chakrabarti, A. (Eds.) *Universals, Concepts and Qualities*. Aldershot: Ashgate, 239–247.

Armstrong, D.M. (2009). Defending Categoricalism. In Bird, A., Ellis, B. & Sankey, H. (Eds.) *Properties, Powers and Structures*. New York: Routledge, 27–33.

Ashwell, L. (2010). Superficial Dispositionalism. *Australasian Journal of Philosophy* 88 (4), 635–53.

Baez, J. (2001). Higher Dimensional Algebra and Planck Scale Physics. In Callender, C. & Huggett, N. (Eds.), *Physics Meets Philosophy at the Planck Scale*. Cambridge: Cambridge University Press, 177–195.

Balashov, Y. (1999). Zero-Value Physical Quantities. *Synthese* 119, 253–286.

Ball, D. (2011). Property Identities and Modal Arguments. *Philosophers' Imprint* 11(13), 1–19.

Barrow, J. (2002). *The Constants of Nature: From Alpha to Omega-The Numbers That Encode the Deepest Secrets of the Universe*. New York: Pantheon Books.

Barrow, J. & Tipler, F. (1986). *The Anthropic Cosmological Principle*. Oxford: Clarendon Press.

Bartels, A. (1996). Modern Essentialism and the Problem of Individuation of Spacetime Points. *Erkenntnis* 45, 25–43.

Bartels, A. (2013). Why Metrical Properties Are Not Powers. *Synthese* 190 (12), 2001–2013.

Baxter, D. (2001). Instantiation as Partial Identity. *Australasian Journal of Philosophy* 79, 449–464.

Belot, G. & Earman, J. (2001). Pre-socratic Quantum Gravity. In Callender, C. & Huggett, N. (Eds.) *Physics Meets Philosophy at the Planck Scale*. Cambridge: Cambridge University Press, 213–255.

Bhaskar, R. (1975). *A Realist Theory of Science*. Leeds: Leeds Books.

Bigaj, T. (2010). Dispositional Monism and the Circularity Objection. *Metaphysica* 11, 39–47.

Bigelow, J. & Pargetter, R. (1989). A Theory of Structural Universals. *Australasian Journal of Philosophy* 67(1), 1–11.

Bigelow, J., Ellis, B. & Lierse, C. (1992). The World as One of a Kind. *British Journal for the Philosophy of Science* 43, 371–388.

Bird, A. (1998). Dispositions and Antidotes. *Philosophical Quarterly* 48, 227–234.

Bird, A. (2004). Antidotes All the Way Down? *Theoria* 19, 259–269.

Bird, A. (2006). Potency and Modality. *Synthese* 149, 491–508.

Bird, A. (2007). *Nature's Metaphysics: Laws and Properties.* Oxford: Clarendon Press.

Bird, A. (2007a). The Regress of Pure Powers? *The Philosophical Quarterly* 57, 513–534.

Bird, A. (2009). Structural Properties Revisited. In Handfield, T. (Ed.). *Dispositions and Causes.* Oxford: Oxford University Press, 215–241.

Bird, A. (2012). Monistic Dispositional Essentialism. In Bird, A., Ellis, B. & Sankey, H. (Eds.) *Properties, Powers and Structures.* New York: Routledge, 35–41.

Black, R. (2000). Against Quidditism. *Australasian Journal of Philosophy* 78(1), 87–104.

Blackburn, S. (1990). Filling in Space. *Analysis* 50, 62–65.

Bolender, J. (2006). Nomic Universals and Particular Causal Relations: Which Are Basic and Which Are Derived? *Philosophia* 34, 405–410.

Bostock, S. (2003). Are All Possible Laws Actual Laws? *Australasian Journal of Philosophy* 81(4), 517–533.

Bostock, S. (2008). In Defence of Pandispositionalism. *Metaphysica* 9(2), 139–157.

Bricker, P. (2008). Concrete Possible Worlds. In Sider, T., Hawthorne, J. & Zimmerman, D. (Eds.) *Contemporary Debates in Metaphysics.* Malden, MA: Blackwell, 111–134.

Butterfield, J. & Bouatta, N. (2015). Renormalization for Philosophers. In Bigaj, T. & Wuthrich, C. (Eds.) *Metaphysics in Contemporary Physics.* Poznan Studies in Philosophy of Science.

Campbell, K. (1990). *Abstract Particulars.* Oxford: Blackwell.

Carnap, R. (1936). Testability and Meaning I. *Philosophy of Science* 3, 420–471.

Carnap, R. (1937). Testability and Meaning II. *Philosophy of Science* 4, 1–40.

Carnap, R. (1956). The Methodological Character of Theoretical Concepts. *Minnesota Studies in the Philosophy of Science* I, 38–76.

Carr, B. (Ed.) (2007). *Universe or Multiverse.* Cambridge: Cambridge University Press.

Cartwright, N. (1999). *The Dappled World: A Study of the Boundaries of Science.* Cambridge: Cambridge University Press.

Cartwright, N. (2009). Causal Laws, Policy Predictions, and the Need for Genuine Powers. In Handfield, T. (Ed.) *Dispositions and Causes.* Oxford: Clarendon Press, 127–157.

Castellani, E. (2002). Symmetry, Quantum Mechanics and Beyond. *Foundations of Science* 7, 181–196.

Chakravartty, A. (2007). *A Metaphysics for Scientific Realism*. Cambridge: Cambridge University Press.

Chakravartty, A. (2007a). Inessential Aristotle: Powers Without Essences. In Groff, R. (Ed.) *Revitalizing Causality: Realism About Causality in Philosophy and Social Science*. London/New York: Routledge, 152–162.

Chakravartty, A. (2013). Realism in the Desert and in the Jungle: Reply to French, Ghins and Psillos. *Erkenntnis* 78(1), 39–58.

Chakravartty, A. (forthcoming). Saving the Scientific Phenomena: What Powers Can and Cannot Do. In Jacobs, J. D. (Ed.), *Putting Powers to Work*. Oxford: Oxford University Press.

Chalmers, A. (1999). Making Sense of Laws of Physics. In Sankey, H. (Ed.) *Causation and Laws of Nature*. Kluwer Academic Publishers, 3–18.

Chalmers, D., Manley, D. & Wasserman, R. (Eds.) (2009). *Metametaphysics*. Oxford: Oxford University Press.

Chisholm, R. (1967). Identity Through Possible Worlds. *Nous* 1, 1–8.

Choi, S. (2005). Do Categorical Ascriptions Entail Counterfactual Conditionals? *The Philosophical Quarterly* 55(220), 495–503.

Choi, S. (2008). Dispositional Properties and Counterfactual Conditionals. *Mind* 117, 795–841.

Choi, S. (2012). Intrinsic Finks and the Dispositional/Categorical Distinction. *Nous* 46(2), 289–325.

Choi, S. (2013). Can Opposing Dispositions Be Co-instantiated? *Erkenntnis* 78(1), 161–82.

Clarke, R. (2008). Intrinsic Finks. *The Philosophical Quarterly* 58, 512–8.

Cohen-Tannoutji, G. (2009). Universal Constants, Standard Models and Fundamental Metrology. *European Physical Journal (Special Topics)* 172, 5–24.

Contessa, G. (2013). Dispositions and Interferences. *Philosophical Studies* 165, 401–419.

Corry, R. (2009). How Is Scientific Analysis Possible? In Handfield, T. (Ed.) *Dispositions and Causes*. Oxford: Clarendon Press, 158–188.

Corry, R. (2011). Can Dispositional Essences Ground the Laws of Nature? *Australasian Journal of Philosophy* 89(2), 263–275.

Crane, T. (Ed.) (1996). *Dispositions: A Debate*. London/New York: Routledge.

Cross, T. (2005). What Is a Disposition? *Synthese* 144, 321–341.

Cross, T. (2012). Goodbye, Humean Supervenience. *Oxford Studies in Metaphysics* 7, 129–153.

Cross, T. (2012a). Recent Work on Dispositions. *Analysis* 72(1), 115–124.

Curiel, E. (2000). The Constraints General Relativity Places on Physicalist Accounts of Causality. *Theoria* 15(1), 33–58.

Curiel, E. (2015). If Metrical Structure Were Not Dynamical, Counterfactuals in General Relativity Would Be Easy. (manuscript)

Denkel, A. (1997). On the Compresence of Tropes. *Philosophy and Phenomenological Research* LVII(3), 599–606.

Dodd, J. (2002). Is Truth Supervenient on Being? *Proceedings of the Aristotelian Society* 102(1), 69–86.

Dretske, F. (1977). Laws of Nature. *Philosophy of Science* 44, 248–268.

Duff, M.J. (2004). Comment on the Time-Variation of Fundamental Constants. ArXiv:hep-th/0208093v3 (11 July 2004).

Duff, M.J., Okun, L.B & Veneziano, G. (2002). Trialogue on the Number of Fundamental Constants. *Journal of High Energy Physics* 03(2002), 023. Also in arXiv: physics/0110060v2 [physics.class-ph] (28 Feb 2002).

Earman, J. (1995). *Bangs, Crunches, Whimpers, and Shrieks: Singularities and Acausalities in Relativistic Spacetimes.* New York: Oxford University Press.

Earman, J. (2004). Laws, Symmetry, and Symmetry Breaking: Invariance, Conservation Principles, and Objectivity. *Philosophy of Science* 71(5), 1227–1241.

Earman, J., Roberts, J. & Smith, S. (2002). Ceteris Paribus Lost. *Erkenntnis* 57, 281–301.

Egan, A. (2004). Second Order Predication and the Metaphysics of Properties. *Australasian Journal of Philosophy* 82, 48–66.

Ellis, B. (2001). *Scientific Essentialism.* New York: Cambridge University Press.

Ellis, B. (2002). *The Philosophy of Nature*. Chesham: Acumen.

Ellis, B. (2005). Universals, the Essential Problem and Categorical Properties. *Ratio* XVIII, 462–472.

Ellis, B. (2005a). Katzav on the Limitations of Dispositionalism. *Analysis* 65(1), 90–92.

Ellis, B. (2009). *The Metaphysics of Scientific Realism*. Durham: Acumen.

Ellis, B. & Lierse, C. (1994). Dispositional Essentialism. *Australasian Journal of Philosophy* 72(1), 27–45.

Ellis, G.F.R. (1979). The Homogeneity of the Universe. *General Relativity and Gravitation* 11(4), 281–289.

Esfeld, M. & Sachse, C. (2011). *Conservative Reductionism.* New York: Routledge.

Everett, A. (2009). Intrinsic Finks, Masks and Mimics. *Erkenntnis* 71, 191–203.

Fales, E. (1993). Are Causal Laws Contingent? In Bacon, J., Campbell, K. & Reinhardt, L. (Eds.) *Ontology, Causality and Mind: Essays in Honour of D.M. Armstrong*. Cambridge: Cambridge University Press, 121–144.

Fine, K. (1994). Essence and Modality. In Tomberlin, J. (Ed.) *Philosophical Perspectives 8: Logic and Language*. Atascadero, CA: Ridgeview, 1–16.

French, S. (2011). Metaphysical Underdetermination: Why Worry? *Synthese* 180(2), 205–221.

French, S. (2013). Semi-Realism, Sociability and Structure. *Erkenntnis* 78(1), 1–18.

French, S. (forthcoming). Doing Away with Dispositions: Powers in the Context of Modern Physics. In Meincke-Spann, A.S. (Ed.) *Dispositionalism: Perspectives from Metaphysics and the Philosophy of Science*. Springer Synthese Library, Springer.

French, S. & McKenzie, K. (2012). Thinking Outside the Toolbox: Towards a More Productive Engagement Between Metaphysics and Philosophy of Physics. *European Journal of Analytic Philosophy* 8(1), 42–59.

Friederich, S. (2014). A Philosophical Look at the Higgs Mechanism. *Journal for General Philosophy of Science* 45, 335–350.

Friedman, M. (1983). *Foundations of Spacetime Theories*. Princeton, New Jersey: Princeton University Press.

Funkhouser, E. (2006). The Determinable-Determinate Relation. *Nous* 40(3), 548–569.

Gundersen, L. (2002). In Defence of the Conditional Account of Dispositions. *Synthese* 130, 389–411.

Guth, A. (1981). Inflationary Universe: A Possible Solution to the Horizon and Flatness Problem. *Physical Review D* 23, 347–356.

Hamermesh, M. (1989). *Group Theory and Its Applications to Physical Problems*. New York: Dover Publications.

Handfield, T. (2008). Unfinkable Dispositions. *Synthese* 160, 297–308.

Harnik, R., Kribs, G. & Perez, G. (2006). A Universe Without Weak Interaction. *Physical Review D* 74(3), 035006.

Harre, R. (1970). Powers. *British Journal for the Philosophy of Science* 21(1), 81–101.

Harre, R. & Madden, E.H. (1975). *Causal Powers: A Theory of Natural Necessity*. Oxford: Blackwell.

Haslanger, S. & Kurtz, R.M. (2006). *Persistence: Contemporary Readings*. Mass.: Bradford Books MIT Press.

Hawking, S. & Ellis, G.F.R. (1973). *The Large Scale Structure of Space-Time*. Cambridge: Cambridge University Press.

Hawley, K. (2001). *How Things Persist.* Oxford: Clarendon Press.

Hawley, K. (2010). Temporal Parts. *The Stanford Encyclopedia of Philosophy* (Winter 2010 Edition), Edward N. Zalta (Ed.), http://plato.stanford.edu/archives/win2010/entries/temporal-parts/

Hawthorne, J. (2001). Causal Structuralism. *Philosophical Perspectives* 15, 361–378.

Heil, J. (2003). *From an Ontological Point of View.* New York: Oxford University Press.

Heil, J. (2004). Properties and Powers. In Zimmerman, D. (Ed.) *Oxford Studies in Metaphysics 1.* Oxford: Clarendon Press, 223–254.

Heil, J. (2007). Precis of *From an Ontological Point of View.* In Romano, G. (Ed.) *Symposium on From an Ontological Point of View by John Heil.* SWIF Philosophy of Mind Review 6(2), 11–21.

Heller, M. (1998). Property Counterparts in Ersatz Worlds. *The Journal of Philosophy* 95, 293–316.

Hendry, R.F. & Rowbottom, D.P. (2009). Dispositional Essentialism and the Necessity of Laws. *Analysis* 69(4), 668–677.

Higgs, P.W. (1966). Spontaneous Symmetry Breakdown Without Massless Bosons. *Physical Review* 145, 1156–1163.

Hoefer, C. (2000). Energy Conservation in GTR. *Studies in History and Philosophy of Modern Physics* 31(2), 187–199.

Holton, R. (1999). Dispositions All the Way Round. *Analysis* 59, 9–14.

Hornsby, J. (2005). Truthmaking Without Truthmaker Entities. In Beebee, H. & Dodd, J. (Eds.) *Truthmakers: The Contemporary Debate.* Oxford: Oxford University Press, 33–47.

Humphreys, P. (2013). Scientific Ontology and Speculative Ontology. In Ross, D., Ladyman, J. & Kincaid, H. (Eds.) *Scientific Metaphysics.* Oxford: Oxford University Press, 51–78.

Hüttemann, A. (1998). Laws and Dispositions. *Philosophy of Science* 65, 121–135.

Hüttemann, A. (2007). Causation, Laws and Dispositions. In Kistler, M. & Gnassounou, B. (Eds.) *Dispositions and Causal Powers.* Hampshire: Ashgate, 207–219.

Ingthorsson, R.D. (2013). Properties: Qualities, Powers or Both? *Dialectica* 67(1), 55–80.

Ingthorsson, R.D. (2012). The Regress of Pure Powers Revisited. *European Journal of Philosophy* 23(3), 529–541.

Jacobs, J. (2011). Powerful Qualities, Not Pure Powers. *The Monist* 94(1), 81–102.

Karakostas, V. (2004). Forms of Quantum Non-separability and Related Philosophical Consequences. *Journal for General Philosophy of Science* 35, 283–312.

Katzav, J. (2004). Dispositions and the Principle of Least Action. *Analysis* 64(3), 206–214.

Katzav, J. (2005). On What Powers Cannot Do. *Dialectica* 59(3), 331–345.

Kibble, T.W.B. (1967). Symmetry Breaking in Non-abelian Gauge Theories. *Physical Review* 155, 1554–1561.

Kistler, M. (2002). The Causal Criterion of Reality and the Necessity of Laws of Nature. *Metaphysica* 3(1), 57–86.

Kripke, S. (1980). *Naming and Necessity*. Cambridge, MA: Harvard University Press.

Kuhlmann, M., Lyre, H. & Wayne, A. (Eds.) (2002). *Ontological Aspects of Quantum Field Theory*. Singapore: World Scientific Publishing Co.

Ladyman, J. & Ross, D. (2007). *Every Thing Must Go*. New York: Oxford University Press.

Lam, V. (2011). Gravitational and Non-gravitational Energy: The Need for Background Structures. *Philosophy of Science* 78(5), 1012–1024.

Lange, M. (2002). *An Introduction to the Philosophy of Physics*. Oxford: Blackwell.

Lehmkuhl, D. (2008). Is Spacetime a Gravitational Field? In Dieks, D. (Ed.) *The Ontology of Spacetime II*. Amsterdam: Elsevier, 83–110.

Levy-Leblond, J.M. (1979). The Importance of Being (a) Constant. In Toraldo di Francia, G. (Ed.) *Problems in the Foundations of Physics*. Enrico Fermi School LXXII. Amsterdam: North Holland Publications, 237–263.

Lewis, D.K. (1986). *On the Plurality of Worlds*. Oxford: Blackwell.

Lewis, D. (1997). Finkish Dispositions. *Philosophical Quarterly* 47, 143–158.

Lewis, D. (2001). Truthmaking and Difference-Making. *Nous* 35(4), 602–15.

Lewis, D. (2009). Ramseyan Humility. In Brandon-Mitchell, D. & Nola, R., (Eds.) *Conceptual Analysis and Philosophical Naturalism*. Cambridge, MA.: MIT Press, 203–222.

Linde, A.D. (1986). Eternally Existing Self-Reproducing Chaotic Inflationary Universe. *Physics Letters B* 175(4), 395–400.

Linsky, B. & Zalta, E.N. (1994). In Defence of the Simplest Quantified Modal Logic. In Tomberlin, J. (Ed.) *Philosophical Perspectives: Logic and Language*. Atascadero: Ridgeview Press, 431–458.

Livanios, V. (2008). Bird and the Dispositional Essentialist Account of Spatiotemporal Relations. *Journal for General Philosophy of Science* 39(2), 383–394.

Livanios, V. (2010). Symmetries, Dispositions and Essences. *Philosophical Studies* 148(2), 295–305.
Livanios, V. (2012a). Exploring the Metaphysics of Nomic Relations. *Acta Analytica* 27, 247–264.
Livanios, V. (2012b). Is There a (Compelling) Gauge-Theoretic Argument Against the Intrinsicality of Fundamental Properties? *European Journal of Analytic Philosophy* 8(2), 30–38.
Livanios, V. (2014). Categorical Structures and the Multiple Realisability Argument. *METHODE* 3(4), 141–166.
Locke, D. (2012). Quidditism Without Quiddities. *Philosophical Studies* 160(3), 345–363.
Loux, M. (1998). *Metaphysics: A Contemporary Introduction*. New York/London: Routledge.
Lowe, J. (2002). *A Survey of Metaphysics*. Oxford: Oxford University Press.
Lowe, J. (2006). *The Four-Category Ontology: A Metaphysical Foundation for Natural Science*. Oxford: Clarendon Press.
Lyre, H. (2008). Does the Higgs Mechanism Exist? *International Studies in the Philosophy of Science* 22(2), 119–133.
Lyre, H. (2012). The Just-So Higgs Story: A Response to Adrian Wuthrich. *Journal for General Philosophy of Science* 43(2), 289–294.
MacBride, F. (1999). Could Armstrong Have Been a Universal? *Mind* 108(431), 471–501.
Mackie P. (2006). *How Things Might Have Been*. Oxford: Clarendon Press.
Malzkorn, W. (2000). Realism, Functionalism and the Conditional Analysis of Dispositions. *The Philosophical Quarterly* 50, 452–69.
Manley, D. (2012). Dispositionality: Beyond the Biconditionals. *Australasian Journal of Philosophy* 90(2), 321–334.
Manley, D. & Wasserman R. (2007). A Gradable Approach to Dispositions. *The Philosophical Quarterly* 57, 68–75.
Manley, D. & Wasserman R. (2008). On Linking Dispositions and Conditionals. *Mind* 117, 59–84.
Martin, C.B. (1993). Power for Realists. In Bacon, J., Campbell, K. & Reinhardt, L. (Eds.) *Ontology, Causality and Mind: Essays in Honour of D.M. Armstrong*. Cambridge: Cambridge University Press, 175–186.
Martin, C.B. (1994). Dispositions and Conditionals. *Philosophical Quarterly* 44, 1–8.
Martin, C.B. (1997). On the Need for Properties: The Road to Pythagoreanism and Back. *Synthese* 112, 193–231.

Martin, C.B. (2008). *The Mind in Nature.* Oxford: Clarendon Press.
Martin, C.B. & Heil, J. (1999). The Ontological Turn. *Midwest Studies in Philosophy* XXIII, 34–60.
McKenzie, K. (2011). Arguing Against Fundamentality. *Studies in History and Philosophy of Modern Physics* 42, 244–255.
McKenzie, K. (2014) In No Categorical Terms: A Sketch for an Alternative Route to Humeanism About Fundamental Laws. In Galavotti, M.C. et al. (Eds.) *New Directions in the Philosophy of Science. The Philosophy of Science in a European Perspective, Vol.5.* New York: Springer, Ch.4.
McKitrick, J. (2003). The Bare Metaphysical Possibility of Bare Dispositions. *Philosophy and Phenomenological Research* 66, 349–69.
McKitrick, J. (2009). Dispositional Pluralism. In Damschen, G., Schnepf, R. & Stueber, K. (Eds.) *Debating Dispositions: Issues in Metaphysics, Epistemology and Philosophy of Mind.* New York/Berlin: De Gruyter, 186–203.
Melia, J. (2005). Truthmaking Without Truthmakers. In Beebee, H. & Dodd, J. (Eds.) *Truthmakers: The Contemporary Debate.* Oxford: Oxford University Press, 67–84.
Mellor, D.H. (1974). In Defence of Dispositions. *Philosophical Review* 83, 157–181.
Mellor, D.H. (1982). Counting Corners Correctly. *Analysis* 42, 96–97.
Mellor, D.H. (2000). The Semantics and Ontology of Dispositions. *Mind* 109(436), 757–780.
Misner, C., Thorne, K. & Wheeler, J.A. (1973). *Gravitation.* San Francisco: W.H. Freeman and Co.
Moffat, J.W. (2002). Comment on the Variation of Fundamental Constants. ArXiv: hep-th/0208109v2 (7 Nov 2002).
Molnar, G. (2003). *Powers: A Study in Metaphysics.* New York: Oxford University Press.
Moreland, J.P. (2001) *Universals.* Chesham: Acumen.
Morganti, M. (2014). Metaphysical Infinitism and the Regress of Being. *Metaphilosophy* 45(2), 232–244.
Mumford, S. (1998). *Dispositions*, New York: Oxford University Press
Mumford, S. (2004). *Laws in Nature.* London: Routledge.
Mumford, S. (2006). The Ungrounded Argument. *Synthese* 149(3), 471–489.
Mumford, S. & Anjum, R.L. (2011). *Getting Causes from Powers.* Oxford: Oxford University Press.
Nerlich, G. (2010). Why Spacetime Is Not a Hidden Cause: A Realist Story. In Petkov (Ed.) *Space, Time, and Spacetime.* Berlin/Heidelberg: Springer Verlag, 181–191.

Noether, E. (1918). Invariante Variationsprobleme. *Nachr. d. Konig. Gesellschd. d. Wiss. Zu Gottingen, Math-phys. Klasse,* 235–57.
O' Connor, T. & Wong, H.Y. (2015). Emergent Properties. *The Stanford Encyclopedia of Philosophy.* (Summer 2015 Edition). Zalta, E.N. http://plato.stanford.edu/archives/sum2015/entries/properties-emergent/
Oderberg, D. (2009). The Non-identity of the Categorical and the Dispositional. *Analysis* 69(4), 677–684.
Oderberg, D. (2011). The World Is Not an Asymmetric Graph. *Analysis* 71(1), 3–10.
Oderberg, D. (2012). Graph Structuralism and Its Discontents: Rejoinder to Shackel. *Analysis* 72(1), 94–98.
Peskin, M.E. & Schroeder, D.V. (1995). *An Introduction to Quantum Field Theory*. Reading, Mass.: Addison-Wesley.
Place, U.T. (1996). Dispositions as Intentional States. In Crane, T. (Ed.) *Dispositions: A Debate*. London/New York: Routledge, 19–32.
Prior, E.W. (1982). The Dispositional/Categorical Distinction. *Analysis* 42, 93–96.
Prior, E.W. (1985). *Dispositions*. Aberdeen: Aberdeen University Press.
Prior, E.W., Pargetter, R. & Jackson, F. (1982). Three Theses About Dispositions. *American Philosophical Quarterly* 19, 251–257.
Psillos, S. (2006). What Do Powers Do When They Are Not Manifested? *Philosophy and Phenomenological Research* LXXII(1), 137–56.
Psillos, S. (2008). Cartwright's Realist Toil: From Entities to Capacities. In Hartmann, S., Hoefer, C. & Bovens, L. (Eds.) *Nancy Cartwright's Philosophy of Science*. New York/London: Routledge, 167–194.
Psillos, S. (2012). Adding Modality to Ontic Structuralism: An Exploration and Critique. In Landry, E. & Rickles, D. (Eds.) *Structural Realism: Structure, Object and Causality*. Dordrecht: Springer, 169–185.
Rees, M. (1999). *Just Six Numbers*. New York: Basic Books.
Roberts, J. (2010). Some Laws of Nature Are Metaphysically Contingent. *Australasian Journal of Philosophy* 88(3), 445–457.
Rodriguez-Pereyra, G. (2006). Truthmakers. *Philosophy Compass* 1/2, 186–200.
Rosenkrantz, G.S. (1993). *Haecceity: An Ontological Essay*. Dordrecht: Kluwer.
Rovelli, C. (2004). *Quantum Gravity.* Cambridge: Cambridge University Press.
Ryder, L. (1985). *Quantum Field Theory*. Cambridge: Cambridge University Press.
Schaffer, J. (2005). Quiddistic Knowledge. *Philosophical Studies* 123, 1–32.
Schneider, S. (2001). Alien Individuals, Alien Universals and Armstrong's Combinatorial Theory of Possibility. *Southern Journal of Philosophy* 39(4), 575–593.

Schrenk, M. (2009). Hic Rhodos, Hic Salta: From Reductionist Semantics to a Realist Ontology of Forceful Dispositions. In Damschen, G., Schnepf, R. & Stueber, K. (Eds.) *Debating Dispositions: Issues in Metaphysics, Epistemology and Philosophy of Mind*. New York/Berlin: De Gruyter, 145–167.
Schrenk, M. (2010). The Powerlessness of Necessity. *Nous* 44(4), 725–739.
Schroer, R. (2010). Is There More Than One Categorical Property? *The Philosophical Quarterly* 60(241), 831–850.
Schroer, R. (2013). Can a Single Property Be Both Dispositional and Categorical? The "Partial Consideration Strategy", Partially Considered. *Metaphysica* 14(1), 63–77.
Shackel, N. (2011). The World as a Graph: Defending Metaphysical Graphical Structuralism. *Analysis* 71(1), 10–21.
Shoemaker, S. (1980). Causality and Properties. In Van Inwagen, P. (Ed.) *Time and Cause*. Dordrecht: Reidel, 109–135.
Sider, T. (2001). *Four-Dimensionalism*. Oxford: Oxford University Press.
Simons, P. (1994). Particulars in Particular Clothing: Three Trope Theories of Substance. *Philosophy and Phenomenological Research* 54(3), 553–575.
Simons, P. (2010). Relations and Truthmaking. *Proceedings of the Aristotelian Society*, Supplementary Volume LXXXIV, 199–213.
Sklar, L. (1976). *Space, Time, and Spacetime*. Berkeley/Los Angeles: University of California Press.
Smart, B. (2013). Categorical Properties in Background Independent Substantivalist General Relativity. http://philpapers.org/rec/SMACPI.
Smeenk, C. (2006). The Elusive Higgs Mechanism. *Philosophy of Science* 73(5), 487–499.
Smith, S. (2002). Violated Laws, Ceteris Paribus Clauses, and Capacities. *Synthese* 130, 235–264.
Smolin, L. (1997). *The Life of the Cosmos*. Oxford: Oxford University Press.
Smolin, L. (2006). *The Trouble with Physics*. Boston/New York: Houghton Mifflin Co.
Stenger, V. (2004). Is the Universe Fine-Tuned for Us?. In Young, M. & Edis T. (Eds.) *Why Intelligent Design Fails: A Scientific Critique of the New Creationism*. New Brunswick, NJ: Rutgers University Press, 172–84.
Strawson, G. (2008). The Identity of the Categorical and the Dispositional. *Analysis* 68(4), 271–282.
Struyve, W. (2011). Gauge Invariant Accounts of the Higgs Mechanism. *Studies in History and Philosophy of Science Part B* 42(4):226–236.

Swinburne, R. (1980). Properties, Causation and Projectibility: Reply to Shoemaker. In Cohen, L. & Hesse, M. (Eds.). *Applications of Inductive Logic*. Oxford: Oxford University Press, 313–320.

Taylor, J. (2013). In Defence of Powerful Qualities. *Metaphysica* 14(1), 93–107.

Tegmark, M., Aguirre, A., Rees, M.J. & Wilczek, F. (2006). Dimensionless Constants, Cosmology, and Other Dark Matters. *Physical Review D* 73(2), 023505.

Teller, P. (1995). *An Interpretive Introduction to Quantum Field Theory*. Princeton, New Jersey: Princeton University Press.

Teller, P. (2002). So What Is the Quantum Field? In Kuhlmann, M., Lyre, H. & Wayne, A. (Eds.). *Ontological Aspects of Quantum Field Theory*. Singapore: World Scientific Publishing Co., 145–162.

Thompson, I.J. (1988). Real Dispositions in the Physical World. *British Journal for the Philosophy of Science* 39, 67–79.

Tooley, M. (1977). The Nature of Laws. *Canadian Journal of Philosophy* 7, 667–698.

Tugby, M. (2012). The Metaphysics of Pan-Dispositionalism. In Bird, A., Ellis, B. & Sankey, H. (Eds.) *Properties, Powers and Structures*. New York: Routledge, 165–179.

Tugby, M. (2013). Platonic Dispositionalism. *Mind* 122(486), 451–480.

Uzan, J.P. (2002). The Fundamental Constants and Their Variation: Observational Status and Theoretical Motivations. *Reviews of Modern Physics* 75(2), 403–455. Also in arXiv: hep ph/0205340v1(30 May 2002).

Uzan, J.P & Leclercq, B. (2008). *The Natural Laws of the Universe: Understanding Fundamental Constants*. Chichester, UK: Praxis Publishing.

Vallentyne, P. (1998). The Nomic Role Account of Carving Reality at the Joints. *Synthese* 115, 171–198.

Vetter, B. (2013). Multi-Track Dispositions. *The Philosophical Quarterly* 63 (251), 330–352.

Vilenkin, A. (1983). Birth of Inflationary Universes. *Physical Review D* 27, 2848–2855.

Wald, R. (1984). *General Relativity*. Chicago/London: The University of Chicago Press.

Weinberg, S. (1992). *Dreams of a Final Theory*. New York: Pantheon Books.

Weinberg, S., Nielsen, H.B. and Taylor, J.G. (1983). Overview of Theoretical Prospects for Understanding the Values of Fundamental Constants. *Philosophical Transactions of the Royal Society of London A* 310, 249–252.

Wheeler, J.A. (1974). Beyond the End of Time. In Rees, M., Ruffini, R. & Wheeler, A.J. (Eds.) *Black Holes, Gravitational Waves and Cosmology*. New York: Gordon and Breach.

Wigner, E. (1939). On Unitary Representations of the Inhomogeneous Lorentz Group. *Annals of Mathematics* 40, 149–204.

Wilczek, F. (2007). Fundamental Constants. ArXiv: 0708.4361v1 [hep-ph] (31 Aug 2007).

Williams, N.E. (2009). The Ungrounded Argument Is Unfounded: A Response to Mumford. *Synthese* 170(1), 7–19.

Wuthrich, A. (2012). Eating Goldstone Bosons in a Phase Transition: A Critical Review of Lyre's Analysis of the Higgs Mechanism. *Journal for General Philosophy of Science* 43(2), 281–287.

Zwiebach, B. (2009). *A First Course in String Theory*. Cambridge: Cambridge University Press.

Index

Note: Page numbers followed by 'n' denote notes

V. Livanios, *Science in Metaphysics*,
DOI 10.1007/978-3-319-41291-7

D

E

CPI Antony Rowe
Chippenham, UK
2017-09-18 09:41